Pedro Albertos
Iven Mareels
(Authors)
Feedback and Control for Everyone

Pedro Albertos
Iven Mareels
(Authors)

Feedback and Control for Everyone

With 158 Figures

 Springer

Authors

Prof. Dr. Pedro Albertos

Department of Systems Engineering and Control
University Politecnica Valencia
POB 22012
46071 Valencia
Spain

Prof. Dr. ir Iven Mareels

Melbourne School of Engineering
The University of Melbourne
Melbourne Vic 3010
Australia

ISBN 978-3-642-03445-9

Library of Congress Control Number: 2009941651

© Springer-Verlag Berlin Heidelberg 2010

This work is subject to copyright. All rights are reserved, whether the whole or part of the material is concerned, specifically the rights of translation, reprinting, reuse of illustrations, recitations, broadcasting, reproduction on microfilm or in any other way, and storage in data banks. Duplication of this publication or parts thereof is permitted only under the provisions of the German Copyright Law of September 9, 1965, in its current version, and permission for use must always be obtained from Springer. Violations are liable to prosecution under the German Copyright Law.

The use of general descriptive names, registered names, trademarks, etc. in this publication does not imply, even in the absence of a specific statement, that such names are exempt from the relevant protective laws and regulations and therefore free for general use.

Cover design: WMX Design GmbH, Heidelberg
Typesetting: Stasch, Bayreuth (stasch@stasch.com)
Production: Heather King
Language editing: Georg Poese, Sariac-Magnoac (France)

Printed on acid-free paper

9 8 7 6 5 4 3 2 1

springer.com

To our friends, teachers and students

Preface

Vertrouwen hebben is goed, in controle zijn is beter!
Confidence is good, but to be in control is better!
La confianza está bien, ¡pero el control es mejor...!

La información suele ser buena, ¡la realimentación es esencial!
Information is important but feedback is essential!
Informatie is belangrijk, maar feedback is essentieel!

Our journey to write a general introduction to control and feedback started with the realisation that in our chosen field of study the proponents have taken little time to project the main ideas outside the community of the initiated. We are constantly reminded of the fact that there is no general awareness of control engineering, or control technology, despite its ubiquity when we try to introduce ourselves as control engineers. Invariably this leads to *A what engineer?* response. We suspect that most control engineers do not even try to introduce themselves as such. It certainly is a poor communication starter.

Colleagues, friends, collaborators often ask rather natural questions like *So, what do you study in control? Is there a non-specialist introduction to the field? Can I read something to get a feel about this field of systems theory? What questions do you try to answer? What are the key results?* In reply we can point to a wealth of technical literature, rooted in mathematics, quite acceptable to some engineers or mathematics graduates, but rather too formidable a hurdle for most. Our apologetic replies, complemented with the odd reference to the history of the field or some systems philosophy, invariably raises eyebrows, and evokes surprise, disbelief and disapproval.

But how do you write about a theory of everything that is applied everywhere? Clearly abstraction is the way, but abstraction with precision is for engineers (like we are) a one way street into mathematics.

The task appears too hard. Certainly we can easily understand why so little has been done in trying to make control theory and control engineering more widely accessible. Indeed, excuses are all too easy to find.

- Someone else, better qualified, should/will do it.
- Our peers will disapprove of our sloppiness, and others may find the material still too forbidding. It is a laudable quest, but one with more opportunity for failure than success.
- We possibly cannot do this justice, the field is just too big, and still too rapidly moving, too immature. The whole idea is better postponed.
- Doing research is more important, there is so much that still needs to be done, and can be done, and at least in doing research we are pushing the boundaries of knowledge.

Yet, with the encouragement of a few friends, and we would like to single out Petar Kokotovic in particular, and the generous support of our families, we finally ran out of excuses and challenged ourselves sufficiently to start writing about control, for everyone. Not because we are convinced that we can do this well, but rather because it does matter. Control theory and control engineering is instrumental in the engineered world, and feedback is as fundamental in our natural world as in our technological world. Feedback and control deserve a larger audience.

How and Where This Book Was Written

Our book project stretches out over 5 years, and progressed like two steps forward, one backward, two steps forward, one backward. Talking with our common friend, Karl J. Åström about the way to reach young people (kids, grandsons, children) with complicated matters, we found a challenging book *The Way Life Works* (Hoagland and Dodson 1995) showing it was possible to transmit key concepts with simple language. The authors of this book use a lot of illustrations, which we were not able to do, but the idea was caught and we started to work on it.

Concentrated periods away from our daily duties (research, teaching, management, the usual stuff, all in the familiar environment) were essential to bringing this book to life. We spent weeks at a time, locked away from friends and family and colleagues. In unfamiliar scholarly scenarios, liberated from the internet and mobile telephony, we could focus on writing, and rewriting, but most importantly criticizing our work: "Of course there is too much mathematics". "No, this is far too sloppy". "You cannot possibly see the subtle points that we ought to make here". "How on earth can anybody understand what we want to say here?". "Clearly this goes far too deep". "Sorry but this is too superficial". "There is no substance left here".

Thus, we needed to constantly challenge ourselves that the book is intended for **everyone**, as it all too easy to slip back into our familiar territory. In this sense, we are grateful to a number of high school students and some of our own second year students who voluntarily read some chapters and provided invaluable *feedback*. As a matter of fact, we introduced additional examples and further explanations after the pre-reading of these young "collaborators". Thanks to Miguel, as well as Sergio, Carmen, Tamara, Pablo, Miguel Angel, Felix, Saul, Ismael, Pura, Alvaro, Mercedes and many others.

In this way, we spent time in Melbourne, Valencia, Prague, Seville, Singapore, San Diego, Athens and finally Melbourne again to put the capstone down. To our friends who hosted us, the most unsociable collaborators ever, we owe a great debt of gratitude. Thank you.

What to Expect

If you expect to find elegant mathematics, unambiguously stating what control is all about, and what its great achievements are, do not continue any further. You can lay the book down right now. We have abandoned any attempt for precision, generality or completeness. Brevity excludes both precision and generality, and our lack of knowledge makes it impossible to be complete. Despite the fact simplicity wins over precision, we never try to mislead either, where precision matters and details lack, it will hopefully be apparent. Our main aim, though, is to provide intuition, and motivation, and convey some of the beauty of the key ideas as well as the impact of the engineering achievements. If we perhaps excite a few to go on and read or study the technical literature, that would be a welcome bonus.

If you expect to find no mathematics at all, sorry we will disappoint you too. We valiantly did try to eliminate most mathematics, but clearly control and systems theory is a mathematical field. Its generality demands all the abstraction mathematics can muster, and to rip it all out, would be to destroy its very heart. The use of mathematics is really limited, and most of the text does not rely on it at all. In a few places we did not shy away from the occasional equation, and mathematical expressions, but most of that can either be skipped or glanced over in a first reading, and normally they are accompanied with text to explain what the equations say. For those, though, who get excited as much as ourselves, and are willing to master the ideas behind signals and systems, or even push its boundaries, some mathematical sophistication and computational expertise will have to become a part of life. The journey is not without effort, but the rewards are substantial.

It is our intention that most of the book should be accessible with a basic understanding of typical high school calculus. Also, hopefully it is sufficiently self-contained to allow the reader who wishes to skip the more mathematical bits, to do so and go on with the narrative without penalty. We will let you be the judge.

Besides ideas, we also want to illustrate through examples some of the technology, what it is capable of, and where it has been applied and where it is headed. After all creating technology is what engineers[1] do.

How to read this book

Essentially our exposition is a narrative using intuition and simple examples to convey what matters, and what does not in this field.

You can read the different chapters almost in a random way. In each one of them you will find (we hope) suggestive ideas without the need of much background. Nevertheless, the ideas have been developed in a particular order that we believe will make it easier to follow the narrative. The first chapter is similar to a summary of the book, the content of which is worked over in the following chapters. Some of the "reviewers" preferred that we first define a concept and then illustrate it through examples and applications. We decided against this early on in the book, in Chap. 3 to be precise, and walked through a number of different examples where the presence of feedback and control is of undisputed importance.

The introductory Chap. 1 develops a feel for the generality of the field, and establishes some common vocabulary. In some sense, most of the ideas and concepts are repeated and largely described afterwards. The examples, used here, are revisited from different angles throughout the remainder of the book. At the end of this chapter, like at the end of every chapter, there is a section highlighting the main points and providing hints for further reading and study.

---

[1] The word engineer comes from the Latin *ingenerare*, a word that can be translated as *to create*.

In Chap. 2, the idea of analogy is developed, both for signals and systems. Analogies are in our opinion at the heart of the abstraction so pervasively used in the study of signals and systems. The aim is to convince everyone that it is actually possible to build a meaningful theory of everything, applicable to anything, as long as the limitations are clear. Analogy is certainly not exclusively the domain of systems theory. In fact, the notion of analogy is rather more common in psychology and the social sciences as a methodology to progressively extend concepts from a simple setting to a more complicated yet in some sense analogous scenario.

In Chap. 3 we describe some processes where feedback and control are essential. The engineered examples we have chosen are a manufacturing process for ceramic tiles, a car wash, a large scale distribution system for irrigation water and a radio astronomy antenna. We concentrate on what is measured, how this information is used to decide what to do next, and how decisions are then executed and determine the future behavior of the process. If sensing, reasoning, computing, deciding and actuation are done automatically, we are dealing with an automatic control system. If control is reacting to measurements, feedback is at work. Many human-made control systems are inspired by nature and feedback is prevalent in all natural processes. In the human body, all homeostasis is feedback-based. Most psychological and social processes directly or indirectly use the advantages of feedback.

The two basic ingredients we need to understand feedback are signals and systems; signals, because feedback must measure before it can take action, and systems because feedback is used to influence their behavior. Moreover, both ingredients are interlaced: whenever we refer to systems we refer to something that can create or modify a signal. In Chap. 4 signals are presented. What is a signal? Which properties do they have? How can we represent them? How do we process them? As usual, we use suggestive examples to communicate the message, but everyone will be able to identify a large number of signals in their direct environment: sounds (music, noise, birds chants), sensations (temperature, humidity, light, smell) and many other physical or physiological quantities (force, speed, displacement, weight, humor, tiredness, concentration). As quantitative reasoning is at the core of feedback, the topic is after all approached from an engineering discipline point of view, the mathematical representation of signals play an important role. Some mathematics is used. Nevertheless, the narrative and the reasoning is never based on mathematical formalism.

Central to systems theory and to feedback is the concept of models. The aim of Chap. 5 is to present the basic ideas and tools used in dealing with systems and their models. A model is a convenient representation of a real process of interest, convenient will often mean "it computes". We mainly consider simple examples to illustrate the power of modeling. We briefly discuss how to arrive at a model and pay some attention to structural properties.

Some basic system properties that matter in a feedback context, namely stability, sensitivity and robustness are presented all too briefly in Chap. 6. These concepts are the subject of much and rigorous research. In fact there is no end to studying systems theory. After all it is the theory of everything, so job security will never be an issue. In particular when dealing with large scale or complex systems the theory is really in its infancy.

Chapter 7 deals with the central topic of the book. At this point in the book, most of the ideas about feedback have already been introduced and here, the advantages and disadvantages of feedback are (re)visited and summarized. It will transpire that feedback can be used as an advantage to change the behavior of systems in a dramatic way, despite the fact that it actually imposes constraints on the realm of possibilities.

It is important to emphasize that some form of control or feedback is at work in any system that performs well. The control subsystem and the possible structures for interaction with control are the topics in Chap. 8. Here you will read about the many different ways in which control can interface with a process, and how it alters the usefulness and capability of control. We will lead you to the conclusion that the best performance can be achieved when process and control are considered together, in an integrated way, from the very beginning of design.

Technology is an important driver in the development of feedback, and in turn feedback is a major driver of technology. The different components that make up the control subsystem, sensors, communication, filters, actuators, computers and their software are presented in Chap. 9. This chapter introduces some of the present hardware and software options available to implement control. No doubt, the content of this chapter will soon belong to the realm of archeology.

Conception, design and testing come next. There are a number of design and validation methods supported by computational tools available to approach control design. Some of them are outlined in Chap. 10. The variety of models, goals, constraints and signals involved in control design make it impossible to enumerate all the options. There is still much more work to be done in this arena. Control design requires a lot of creativity and ingenuity, and there is no single recipe for success. Moreover, despite the fact we indicated that the correct approach for control is to conceive it from the onset as part of the problem of overall system design, there are actually no rigorous developments in engineering design along this route (at least as far as we know).

Obviously, we believe that feedback and control matter a lot, and indeed offer a lot to society. In Chap. 11, some of these benefits are highlighted. We hope that you will identify many other benefits. In fact, the purpose of the whole book is to lead you to realize how much can be obtained by means of an appropriate use of feedback and control, and how much has been implemented, so that this *hidden technology*[2] will become clearly exposed. To be balanced, we do also point out some of the risks associated with feedback, but when you get this far through the book you will be well-equipped to decide on such matters for yourself.

Our journey into control concludes in Chap. 12 with a look towards the unpredictable future. Some trends are clear. Some immediate evolutionary technology is easily guessed, but we also present some bolder predictions. The future is open and we will be excited if some of our readers decide to contribute to expand further on the *joy of feedback*[3].

---

[2] Directly quoting Karl Astrom, an excellent teacher, researcher and practitioner and a great advocate for control. We always enjoy to be in touch with him.
[3] Quoting Petar Kokotovic, a friend, an inspiration and a great leader in control and feedback.

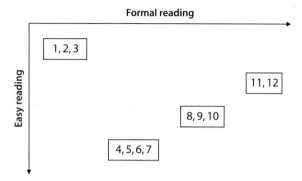

Reading paths. You can follow the formal path, reading chapters sequentially as presented, or you can follow an easier path: start from 1 to 3 and jump to the last two chapters, to get a more informal flavor. Then, go to Chap. 8 to 10 where some mathematics ar used but are still very simple. Finally, Chap. 4 to 7 introduce some concepts where mathematics are "unavoidable" to better understand their meaning.

By the way, each chapter is illustrated by a picture trying to summarize the essence of the concepts of that chapter. We are most grateful to Arturo for his excellent work in capturing in a simple image our engineering lingo.

Pedro and Iven

October 2009
Melbourne

The authors with two sources of inspiration: The Athens' Parthenon (2008) and a *sinusoidal* column in Prague (2005).

Contents

1 Introduction	1
1.1 Feedback	4
1.2 Block Diagrams, Systems, Inputs and Outputs	5
1.3 More on Block Diagrams	8
1.3.1 Recording Music	9
1.3.2 Connected Water Tanks	10
1.3.3 Summary Block Diagrams	11
1.4 Feedback and Dynamics	13
1.4.1 Feedback in the Shower	13
1.4.2 Acoustic Feedback	14
1.4.3 Boiler Water Level Control	14
1.4.4 Birth and Death Processes	15
1.4.5 Manufacturing and Robots	15
1.4.6 Feedback Design and Synthesis	16
1.5 Systems, Causality, Stationarity and Linearity	16
1.6 Models	19
1.6.1 Modeling	21
1.6.2 System Interconnection	22
1.7 The Basic Control Loop	22
1.8 Control Design	24
1.9 Concluding Remarks	25
1.10 Comments and Further Reading	26
2 Analogies	29
2.1 Introduction and Motivation	31
2.2 Signals and Their Graphical Representation	32
2.3 Signals and Analogies	36
2.4 Example Systems	38
2.4.1 Heating an Oven	38
2.4.2 Filling the Kitchen Sink	40
2.4.3 Charging a Capacitor	40
2.4.4 Computer Algorithm	41
2.5 How Similar Is the Behavior of These Example Systems?	43
2.6 Discrete Time or Continuous Time?	43

2.7	Analogous Systems	44
2.8	Combining Systems and Splitting Distributed Systems	46
2.9	Oscillations	47
	2.9.1 Energy Exchange	47
	2.9.2 Systems Perspective	47
2.10	Comments and Further Reading	49

3 Illustrating Feedback and Control ... 51
3.1 Introduction and Motivation ... 53
3.2 Manufacturing Ceramic Tiles ... 54
3.3 Gravity-Fed Irrigation Systems ... 59
3.4 Servo Design for Antennae for Radio Astronomy ... 65
3.5 Simple Automata ... 71
3.6 Homeostasis ... 72
3.7 Social Systems ... 76
    3.7.1 Simple Control Structures ... 79
    3.7.2 Other Control Approaches ... 82
3.8 Comments and Further Reading ... 84

4 Signal Analysis ... 85
4.1 Introduction and Motivation ... 87
4.2 Signals and Signal Classes ... 88
    4.2.1 Signals Defined through Mathematical Algorithms ... 89
    4.2.2 Periodic Signals ... 95
    4.2.3 Stochastic Signals ... 98
    4.2.4 Chaotic Signals ... 99
4.3 Signal Transforms ... 100
4.4 Measuring a Signal ... 101
    4.4.1 The Size of a Signal ... 102
    4.4.2 Sampling a Signal ... 103
    4.4.3 Sampling Periodic Signals: the Aliasing Problem ... 104
4.5 Signal Processing ... 106
4.6 Recording and Reproduction ... 107
    4.6.1 Speech Recording and Reproduction ... 109
4.7 Comments and Further Reading ... 111

5 Systems and Models ... 113
5.1 Introduction and Motivation ... 115
5.2 Systems and Their Models ... 116
    5.2.1 From Signal to System Model ... 116
    5.2.2 From System to Model ... 118
    5.2.3 Model Classes ... 119
5.3 Interconnecting Systems ... 121
5.4 Simplifying Assumptions ... 123
5.5 Some Basic Systems ... 125
    5.5.1 A Linear Gain ... 125

	5.5.2	Transport Delay	127
	5.5.3	An Integrator	128
	5.5.4	An Integrator in a Feedback Loop	130
5.6	Linear Systems	130
	5.6.1	Linear Operators	131
	5.6.2	A Block Diagram Calculus for Linear Systems	133
5.7	System Analysis	135
	5.7.1	Time Response	135
	5.7.2	Frequency Domain	136
	5.7.3	Cascade Connected Systems	139
	5.7.4	Integrators in Series with Feedback	139
5.8	Synthesis of Linear Systems	140
5.9	State Space Description of Linear Systems	143
5.10	A Few Words about Discrete Time Systems	144
5.11	Nonlinear Models	145
5.12	Comments and Further Reading	146

6 Stability, Sensitivity and Robustness 149
6.1	Introduction and Motivation	151
6.2	Some Examples	152
6.3	Stability of Autonomous Systems	154
6.4	Linear Autonomous Systems	156
	6.4.1 General Linear Autonomous Systems, Discrete Time	157
	6.4.2 Continuous Time, Linear System	158
	6.4.3 Exploring Stability	159
6.5	Nonlinear Systems: Lyapunov Stability	163
	6.5.1 Lyapunov's First Method	163
	6.5.2 Energy and Stability: Lyapunov's Second Method	164
6.6	Non-Autonomous Systems	166
	6.6.1 Linear Systems	166
	6.6.2 Nonlinear Systems	167
	6.6.3 Input-to-State Stability and Cascades	169
6.7	Beyond Equilibria	170
	6.7.1 Limit Cycles and Chaos	171
6.8	Sensitivity	172
	6.8.1 Robustness	174
	6.8.2 Sensitivity Computation	175
	6.8.3 General Approach	176
	6.8.4 Sensitivity with Respect to System Dynamics Variations	176
	6.8.5 Sensitivity Measurements	177
6.9	Comments and Further Reading	178

7 Feedback 179
7.1	Introduction and Motivation	181
7.2	Internal Feedback	182
7.3	Feedback and Model Uncertainties	184

7.4	System Stabilization and Regulation		186
	7.4.1	ISS and Feedback Systems	188
	7.4.2	Linear Feedback Systems	189
	7.4.3	The Nyquist Stability Criterion	191
	7.4.4	Integrator with Delay and Negative Feedback	192
7.5	Disturbance Rejection		193
	7.5.1	Noise Feedback	194
7.6	Two-Degrees-of-Freedom Control		195
7.7	Feedback Design		196
7.8	Discussion		197
7.9	Comments and Further Reading		198

8	**The Control Subsystem**		**201**
8.1	Introduction and Motivation		203
8.2	Information Flow		206
8.3	Control Goals		207
8.4	Open-Loop		210
8.5	Closed-Loop		214
8.6	Other Control Structures		215
	8.6.1	2DoF Control	216
	8.6.2	Cascade Control	216
	8.6.3	Selective Control	217
	8.6.4	Inverse Response Systems	218
8.7	Distributed and Hierarchical Control		219
8.8	Integrated Process and Control Design		221
	8.8.1	Scaling the Process and Its Control	221
	8.8.2	Process Redesign	222
8.9	Comments and Further Reading		223

9	**Control Subsystem Components**		**225**
9.1	Introduction and Motivation		227
9.2	Sensors and Data Acquisition Systems		230
	9.2.1	Transducers	232
	9.2.2	Soft Sensors	233
	9.2.3	Communication and Networking	235
	9.2.4	Sensor and Actuator Networks	235
9.3	The Controller		236
	9.3.1	Automata and PLC	236
	9.3.2	On-off Control	237
	9.3.3	Continuous Control: PID	238
9.4	Computer-Based Controllers		239
9.5	Actuators		241
	9.5.1	Smart Actuators	243
	9.5.2	Dual Actuators	243
9.6	Comments and Further Reading		244

10 Control Design 247
10.1 Introduction and Motivation 249
10.2 Control Design 253
10.3 Local Control 255
    10.3.1 Logic and Event-Based Control 255
    10.3.2 Tracking and Regulation 257
    10.3.3 Interactions 261
10.4 Adaptation and Learning 263
    10.4.1 MRAS: MIT Rule 264
    10.4.2 Self Tuning 265
    10.4.3 Gain Scheduling 266
    10.4.4 Learning Systems 267
10.5 Supervision 269
10.6 Optimization 270
    10.6.1 Controlling the Read/Write Head of a Hard Disk Drive 271
    10.6.2 Model Predictive Control 272
10.7 General Remarks 273
10.8 Comments and Further Reading 274

11 Control Benefits 275
11.1 Introduction 277
11.2 Medical Applications 278
11.3 Industrial Applications 280
    11.3.1 Safety and Reliability 281
    11.3.2 Energy, Material or Economic Efficiency 282
    11.3.3 Sustainability 283
    11.3.4 Better Use of Infrastructure 283
    11.3.5 Enabling Behavior 284
    11.3.6 Some Application Areas 285
11.4 Societal Risks 286
11.5 Comments and Further Reading 286

12 A Look at the Future 289
12.1 Introduction 291
12.2 From Analog Controllers
    to Distributed and Networked Control 291
    12.2.1 Embedded Control Systems 292
    12.2.2 Networked Control Systems 294
    12.2.3 Cyber-Physical Systems 295
12.3 From Automatic Manipulators to Humanoid Robots 296
    12.3.1 The Humanoid Challenge 297
    12.3.2 Master-Slave Systems 298
12.4 Artificial Intelligence in Control 298
    12.4.1 Ambient Intelligence 300
    12.4.2 Agents 301

12.5 Systems and Biology 302
   12.5.1 Bio-Systems Modeling 302
   12.5.2 Biomimetics 303
   12.5.3 Bionics 303
   12.5.4 Bio-Component Systems 304
   12.5.5 Protein, and Nano-Scale Biochemical Engineering 304
12.6 Comments and Further Reading 305

References 307

Index 313

Chapter 1 Introduction

1.1 Feedback
1.2 Block Diagrams, Systems, Inputs and Outputs
1.3 More on Block Diagrams
1.4 Feedback and Dynamics
1.5 Systems, Causality, Stationarity and Linearity
1.6 Models
1.7 The Basic Control Loop
1.8 Control Design
1.9 Concluding Remarks
1.10 Comments and Further Reading

Chapter 1

Introduction

Open a window and look at nature.
You may not see it but control is there.

Control and feedback are truly everywhere and touch all of us. They play a crucial role in both the natural and the engineered world. In fact neither nature nor our engineered world could possibly function without feedback. Yet, despite its prevalence we are often totally unaware of its presence. It is often said that control is a hidden technology, which is not such a bad thing for a technology, however it is not very good for the self-esteem of its engineers.

Intuitively, feedback is usually understood as receiving comments or suggestions about an action we have performed. In our scenario, feedback is information obtained from a system[1] used to change its behavior. Feedback is at the heart of

- *a thermostat*, a device that regulates or controls the temperature in a room or house using the central heating and/or cooling unit. Thermostats are also found in most cars, ovens, freezers and refrigerators, and of course hot water units.
- *a toilet flush*, a control mechanism that ensures that the toilet gets flushed, and that the toilet cistern is automatically refilled to a set level. Similar mechanisms are used in irrigation canals, and wherever a fluid level needs to be regulated.
- *a cruise controller*, a device that more or less keeps the car traveling at a constant speed as set by the driver, despite variations in the inclination of the road. More advanced, but in a similar vein, is the *auto pilot* used in aircraft to fly from way point to way point and the *auto rudder* used in ships to maintain a given heading. More generally, the functionality of modern cars and jetliners, including their safety, critically depends on control.
- the *Copenhagen metro*, a light rail network where the cars are automatically and centrally controlled without the need for a driver in each train.

The following anecdote may serve to illustrate just how hidden control technology is. In the early eighties it was quite common to thank the pilot for a smooth landing with applause. One morning flying into Paris, the pilot was thanked in the usual way for an outstanding landing. Promptly the captain came on the public address system not to thank the passengers but to inform them that it was the auto-pilot that had performed the landing. The applause was well-deserved.

---

[1] This term will become clear, but for the time being, a system is just a collective term to denote an instance of either a process, a physical machine or an instrument, anything that can process information in one form or another.

In nature, feedback is at the core of

- *homeostasis*[2], a term used to refer how in living organisms critically important variables are kept within acceptable limits; such as our body temperature or our blood glucose level for example.
- the Earth's *hydrocycle*; how water evaporates, then condenses into clouds, moves around the globe, precipitates to the Earth, and flows back to the oceans, its main reservoir.
- *prey-predator* behavior, or how a predator species feeds on its prey and therefore depends on that prey's survival for its own survival. Most ecologies depend on feedback to maintain the balance of the species in the ecology.

In nature, just like in the engineered world, control and feedback are essential to maintain life just the way we know it (Hoagland and Dodson 1995). From single cells to the complex ecology of our entire planet, from the grandfather clock to a modern jet liner feedback plays a critical and important role in maintaining functionality.

In this chapter we introduce some of the ideas this book will deal with. A central issue is the notion of feedback and that is where we start. All the concepts are introduced in an intuitive manner, without being very precise, but they are here to make certain that we establish a common (technical) vocabulary. Each concept is explored further in later chapters, and we provide references to go well beyond the material of this book.

1.1 Feedback

Probably the first meaning that comes to mind for feedback is feedback as in *giving or receiving feedback*. Parents provide feedback to their children in order to achieve a desired behavioral outcome. As consumers we are constantly asked to provide our opinion on the level of service we have received or in how much we are (dis)satisfied with a particular product. Medical doctors provide feedback to their patients to assist the natural healing processes. In a university, students are asked to provide feedback to the professors about the quality of the instruction they have received. Presumably, the feedback will allow the teachers to improve the course offering and achieve better student learning outcomes as well as student satisfaction in the next course implementation.

What is important in this context is the direction of the flow of information. In order to provide feedback on teaching, the students have to observe or experience the teacher first. This experience is then communicated, *fed back* to the teacher. The teacher then internalizes the information, and uses it to adjust the course material and presentation style for the next batch of students. The presence of a *directed information loop* is the defining characteristic of feedback; without the closure of the information loop there is no feedback.

The direction associated with the information flow is of course related to *causality*. First comes the experience of the teaching, then this experience is communicated to the teacher, next the teacher interprets this and reacts with improved lectures. Somewhat

---

[2] From Greek, literary to keep the same state.

unflatteringly expressed, feedback is always re-active and hence it is always too late. An event or error or action has to occur before feedback can lead to a re-action. Despite this, feedback is immensely powerful and is typically of central importance in understanding any behavior where feedback occurs, and that is almost every behavior we can think of.

1.2 Block Diagrams, Systems, Inputs and Outputs

We will find it very useful to visualize the information loop associated with feedback in a so-called *block diagram*. We will make extensive use of block diagrams throughout the book.

For the student-teacher example we just discussed, we may use the block diagram as represented in Fig. 1.1. Both the teacher and the student are represented by a separate box. The defining feature of a *box* in a block diagram is that a box can take actions (e.g. the teacher provides lectures, results, and assessment tasks). Also, when we observe a box, or take measurements from it, we obtain information about its possible actions (e.g. we may observe the teacher's command of how to project his/her voice during lectures). Both, actions and observations are indicated by an arrow leaving a box. The arrow leaving a box indicates that information has been extracted from it. We refer to these as *outputs*. Further, a box can receive information, or an *input* and this is indicated by an arrow directed into it. In particular, a box may produce an action in response to an input; but not all of the output has to be initiated by or be in response to an input. The teacher may deliver a marked assignment in response to the received assignment, but most of the lecture material is not in response to student input, although some invariably is. An output of one box can be an input into another box, as is clearly the case in Fig. 1.1.

When we identify a closed path or a cycle in the block diagram, we know that feedback is present.

The boxes in a block diagram are normally referred to as *systems*, or *subsystems*, and the entire block diagram is interpreted as the system of interest.

Block diagrams are a very useful tool to visualize a system, or a part of a system of interest. They provide us with a window, a point of view on a system. Any one system of interest may have a few different block diagrams associated with it, depending on what information we may be interested in, or which signals we want to include. A block diagram captures and communicates to the viewer how (sub)systems interact as well as what we find important.

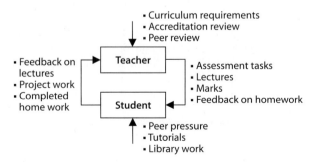

Fig. 1.1. Feedback loop: student-teacher interactions

The overall behavior of the teacher-student interaction will also depend on other information, coming from the environment in which the student-teacher interaction takes place. The teacher will need to plan the teaching in order to meet the curriculum requirements, and will need to satisfy the accreditation board that the expected standards are met. Likewise peer pressure and self study time will play an important role in how much progress the student makes in the class. This other information is also indicated in the block diagram Fig. 1.1, by arrows coming from nowhere into the blocks. These signify external signals, coming from other systems not identified, and loosely referred to as the environment. At times it may be useful to indicate information extracted from the system that is available to the system's environment. Such is represented by arrows leaving a box pointing to nowhere in particular. So it may be perceived that the block diagram in Fig. 1.1 is but a small subset of a potentially much larger block diagram considering many other interactions, and further feedback loops.

Consider as another example Fig. 1.2 which captures the feedback that is present in a generic toilet flush system.

Here we have four boxes or (sub)systems that we focus on: the inlet valve, the water cistern, the outlet valve and the float. To be more complete we could have also considered the water supply, the water bowl and the drain. How the outlet valve is operated has been left out of the picture, it could be either manually operated or by means of a proximity sensor. What we measure, or where we draw the boundary for the block diagram is really our choice, it simply reflects the variables we are interested in. Here,

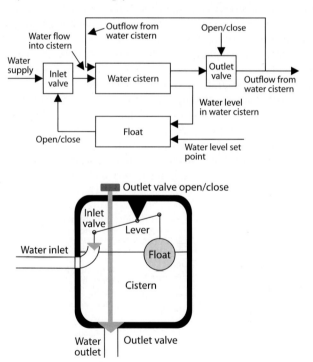

Fig. 1.2. Feedback loop: a simple toilet flushing mechanism

in Fig. 1.2, we are interested in the water level in the water cistern, the inflow of water and the outflow of water as well as the input given to the inlet valve from the float. The inlet flow into the cistern is determined from the valve position and the available water pressure, but we have chosen not to consider these aspects. We simply treat the inlet flow as an input to the cistern as modulated by the valve. The feedback is executed through the float. The float ensures that as long as the water level is below the desired water level the inlet valve remains open. The details of how the inlet valve operates, the mechanics of the float and how the outlet flow comes about could be of interest, but to understand the overall working of the flush system they are actually of little concern.

There is also a second feedback loop inherent in the operation of the cistern: the feedback from the outflow to the cistern. Indeed the outflow also determines the water level in the cistern. This loop nicely illustrates the fact that the arrows in a block diagram are about information, not material flow. If it were only the latter, clearly the only arrow would be out of the cistern as outflow means just that, water leaving the cistern. Nevertheless, the amount of water leaving the cistern determines together with the inflow and the cistern shape the water level in the cistern; the information associated with the outflow is required to determine the water level, hence the arrow from the outlet valve back into the cistern.

The external inputs to the system are the water supply, assumed amply available, the position of the outlet valve and the reference level, which is usually hidden in the input valve mechanism. This is one of the important points in the use of block diagrams. Precise knowledge of the internal workings of the separate subsystems in a block diagram is (often) not required in understanding the overall working of the system captured in the block diagram.

A simple representation of the hydrocycle, illustrated in Fig. 1.3, reveals more generally how recycling in a material flow, not just water, implies feedback. In depicting a block diagram for the hydrocycle we have distinguished three major water reservoirs, the atmosphere, the oceans and the soil; as well as the water flows between them. This

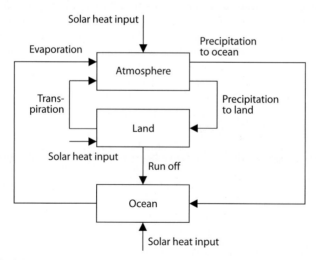

Fig. 1.3. Feedback loop: the hydrocycle or how water is recycled on planet Earth

block diagram reflects that we are interested in water storage and water transport. Any one box stores water, its inputs represent inflows, and its outputs represent water outflows. Clearly a lot is missing (for one, no amount of information about the flows will reveal to us how much water there is actually in the reservoirs). The main energy source driving all of this comes from the Sun, and this is indicated. To make the description more complete we could further subdivide the land reservoir and consider both surface water and ground water.

Collectively, inputs and outputs are referred to as *signals*. Signals vary with time, they are functions of time. A block diagram therefore conveys primarily the idea that a system acts on an input signal to produce an output signal (but remember this is not the only way in which outputs arise). In mathematics this is often referred to as an *operator*. A convenient shorthand for this, (when we do not want to draw a picture) is to write

$$y = P(u) \quad \text{or even simpler} \quad y = Pu$$

This must be read or interpreted as follows: the signal y is the output of the system, represented by the letter P, acting on the input signal, represented by the letter u. The convention is to use lower case letters for signals, and uppercase letters for operators.

For example, we have precipitation (over land and ocean) (p_ℓ, p_o) is a consequence of the atmosphere A acting on evaporation e and transpiration x using the solar heat input, say s. Similarly transpiration and run off r is a consequence of the land mass L acting on the precipitation. Evaporation is the consequence of the ocean acting on precipitation over the ocean and run off from the land mass. Including the solar heat input, which affects all reservoirs, we have six signals e, p_ℓ, p_o, r, x and s constrained by three systems A, L and O^3. All the *signals are functions of time*; e.g. the amount of precipitation changes from minute to minute. To emphasize this dependency, sometimes we will write $y(t)$. To be precise we need to be careful with how the signals are defined, and explain the recipes involved in the operators.

Some care has to be taken using this notation, as systems can produce outputs without the need for an input, and more often than not there may be many different outputs for the same input, depending on the internal condition of the system. In the toilet example the external inputs were summarized, in the hydrocycle only the solar heat input is identified. There are many more signals that we have not included in the block diagram. Deciding on what are the relevant signals in constructing a block diagram depends on the purpose of the block diagram and the level of detail that is required.

1.3 More on Block Diagrams

Signals represent information. In systems theory we like to represent signals and systems in diagrams as in Fig. 1.3. Signals are represented by directed arrows. Systems transform signals in some way and in block diagrams they are (mostly) represented by rectangles.

---

[3] Using this convention, the block diagram of Fig. 1.3 can be written down as follows: $(p_\ell, p_o) = A(e, x, s)$, $(r, x) = L(p_\ell, s)$, and $e = O(p_o, r, s)$.

1.3.1 Recording Music

When recording music through a microphone (Fig. 2.2) all sources of audible sound (a pressure wave in the air that elastically deforms air) in the environment will contribute to the signal that is actually being recorded through the microphone[4]. There is thus an inevitable difference between the signal of interest (the music) and the actual recording which is the signal after the microphone. To capture these differences a number of words are used to distinguish the various signals. The music would be referred to as *the* signal, the other sounds and the artifacts introduced by the microphone are referred to as *disturbances* or simply *noise*. A very simple mathematical representation or *model* for the measured signal is to view it as a sum of the music signal and the noise, like in Fig. 1.4.

Signals are represented by directed arrows. Systems or models transform signals in some way and are represented by rectangles. When the transformation is but a simple sum of signals, or a product we use a circle, in which the symbol $+$ or $\times$ is centered. The arrows that go with the signal indicate the flow of information in the picture. The arrows make the notion of input and output for a particular block intuitively clear, the input is a signal on which the system acts to produce the output signal. In the above situation the flow of information is quite clear, the music sound and the noise are both inputs, whereas the recorded sound is the output.

Another, more complete way of viewing the recording is to consider the whole recording process as a system, where the music together with the other noises are inputs and the measured signal is the output.

In Fig. 1.5, in the block with the label Perform, the signal is transformed from a discrete or digital (D) representation to an analogue one (A). This operation is denoted as Digital to Analogue (D/A) conversion. Here this is performed in a complex manner by a musician. The block labeled Recorder executes the reverse process, the analogue electrical signal and output from the microphone is transformed into a digital signal that is recorded onto the disk. This is an Analogue to Digital (A/D) converter.

Inputs are *free* variables in that they are not restricted by the system to which they are inputs. There is a choice of sheet music. The ambient noise can vary. The recording system does not restrict the music or the noise, both are inputs to it. Output signals are

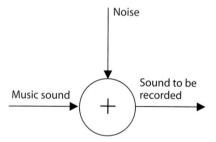

Fig. 1.4. Music corrupted by noise

---

[4] Also, the microphone will not record all sound, as a minimum amount of energy is required in order for the sound to be registered.

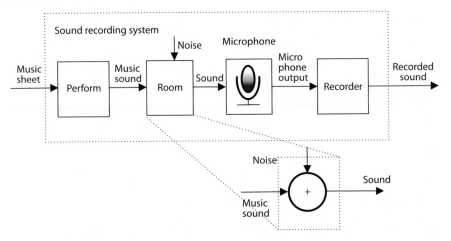

Fig. 1.5. Music corrupted by noise recorded with a microphone

determined by the input signal(s) and the system. Once the music and the noise are given, the microphone generates a particular electrical output signal. There is no freedom left in this output signal.

Of course, the input to one block may be the output of another. So when we say that an input is a free variable, it must be clearly understood that the freedom is with respect to a particular system under consideration.

For example the summation subsystem in Fig. 1.5 does not restrict the noise, nor the music signal, both are inputs to the summation subsystem. The music sound is determined as the output of the digital to analogue block, which takes the sheet music as input and produces the music as output. Similarly the electrical output signal from the microphone is input to the A/D converter that records the digital signal onto the disk. The A/D converter should not restrict in any way the electrical output of the microphone.

The block diagram in Fig. 1.5 reveals a hierarchy of systems, depending on the signals we are interested in. If we do not care for the noisy sound, or the electrical output from the microphone, we have a system that relates the sheet music and the ambient noise to the recorded sound. The hierarchy of systems, the zooming in and out in the picture that is so displayed in the block diagram is one of the properties of block diagrams that make them such a convenient tool for thinking about and communicating system ideas. It is of course possible to zoom in even further, and for example we could unpack what system a microphone is in its own right.

1.3.2 Connected Water Tanks

Consider two water tanks arranged as represented in Fig. 1.6a. The net in(put) flow into the first tank (which is the difference between the inlet flow f_i and outlet flow f_o) will determine the water level in the tank h, as measured from the center of the outlet opening. It is this level that determines the flow out of the tank, the higher the level the higher the flow will be.

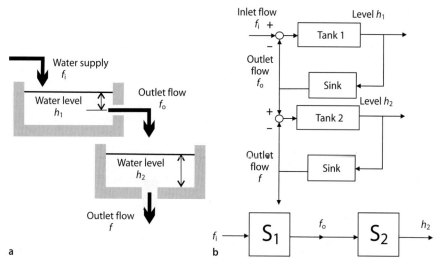

Fig. 1.6. Two tanks: **a** physical schema, **b** block diagram

Similarly, the water level in the second tank is determined by the difference between its inflow (which is the outflow from tank 1) and its outflow. The water level h_2 in the second tank determines the outflow from the tank.

A block diagram representing this situation is depicted in Fig. 1.6b. Notice the feedback in the block diagram.

It is clear that the second tank is not influencing the first, but its behavior depends of course strongly on the first tank. This is an example of a *cascade* of systems, which is reflected in the block diagram by the fact that the respective block diagrams of the subsystems (tank 1 and tank 2) are in *series*: the output of the first system determining the input of the second system.

1.3.3 Summary Block Diagrams

In general, block diagrams are composed of

- (labeled) blocks representing (sub)systems, that act on signals (inputs) and produce signals (outputs);
- directed lines, representing the direction of information flow; these are labeled by the signals.

It is the tradition in drawing block diagrams that some special, frequently occurring blocks that perform very simple operations, are often represented by other than rectangular boxes. Some examples are:

- a *summation block*, typically represented by a circle with a plus sign inside. A summation block's output signal is the sum of all the input signals. The input arrows into the summation block often carry a label + or −, if the label is + (or when no

such label is provided) the input signal gets simply added, if the label is $-$, then the negative of the signal is added to form the output.
- a *multiplication block*, also called a multiplier, typically represented by a circle with a multiplication sign inside. A multiplier is a block whose output signal is the product of its input signals. Input arrows into a multiplier can be labeled with $+$ or $-$ to indicate if the signal or its negative respectively is used in the multiplication.
- a *gain block* or amplifier, typically represented by a triangle. The gain block takes the input signal and multiplies it with a constant, the so-called gain of the amplifier.

Any other operation involving one or several inputs and providing one or several outputs is usually described inside a rectangular box. If the output y is generated as the result of the effect of two inputs, u_1 and u_2, through the operation represented as G, the notation will be $y = G(u_1, u_2)$.

We may identify the following special classes of signals:

- *external input signals*, an arrow into the block diagram, but which does not originate from the block diagram under consideration;
- *external output signal*, an arrow away from the block diagram, but which does not terminate into the block diagram under consideration;
- *internal signal*, an arrow whose endpoints are connected to subsystems in the block diagram.

External signals are said to connect to, or originate from the *environment* of the block diagram. Internal signals are those signals whose cause (origin) and effect (consequence) are linked to the systems in the block diagram.

Signals can connect to more than one block. When this occurs, we say that the directed line representing the signal *bifurcates*.

It is possible for a signal to be both an internal as well as an external output signal.

Some more examples of block diagrams are given in Fig. 1.7.

In Fig. 1.7a, there are three blocks connected in series, from left to right. The block K with external input u and with output connected into a summation which also has the

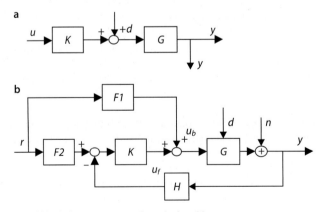

Fig. 1.7. Examples of block diagrams: **a** open loop; **b** closed loop

external input d, the output from the summation leads to the block labeled G, with external output signal y. The output y is bifurcated, carrying the same information to two different elements in the environment, neither of which are shown in the figure. In equation form, we may write that $y = G(d + Ku)$ or in words, y is the output of block G which acts on the input signal $Ku + d$.

In Fig. 1.7b, there are three external input signals r, d and n, there are eight subsystems, five blocks labeled as F_1, F_2, K, G, H and three summation blocks. The signals r and y are bifurcated. The output signal y is both an internal and external signal. To write the relationships expressed by the block diagram in equation format, we consider the summation points, from left to right: $u_f = K(F_2 r - Hy)$, and $y = G(u_f + u_b, d) + n$, where $u_h = F_1 r$ or in words u_f is the output of block K acting on the signal $F_2 r - Hy$, y is the sum of the external signal n and the output of block G acting on the combined signal $(u_f + u_b, d)$.

1.4 Feedback and Dynamics

Feedback and causality are intrinsically connected and therefore time plays an important role in understanding how feedback really works or does not work. Our earlier examples may have given the impression that feedback is easy and always has a positive effect, but this is certainly not the case.

1.4.1 Feedback in the Shower

By way of example consider how we struggle to find the right water temperature in a shower with a manually operated mixing tap (see cover figure in Chap. 7). The best strategy is one of carefully selecting a mixing position, and to wait until the water reaches a relatively steady temperature before adjusting the tap(s) again. Rapid adjustments (feedback) lead invariably to considerable water temperature oscillations and showering discomfort.

What is the problem? Transport delay is the issue. Our correction of the mixing tap(s) position depends on us experiencing the present water temperature. The water temperature on our skin is however quite different from the temperature of the water at the mixing valve due to the time it takes for the water to travel from the mixing valve to our skin. Indeed under typical flow conditions for shower heads the travel time from mixing valve to skin is much longer than the time it takes for us to react to an unpleasant water temperature. If we quickly open up the hot water tap when the water is experienced as too cold, we will then experience the hot water after a minor delay. Reacting, we quickly turn the hot water tap down, only to receive freezing cold water a little later, and so on. Fast feedback creates unpleasant temperature oscillations. The solution is to slow down and to only incrementally adjust the mixing valve.

By the way, a thermostatically controlled mixer tap does not experience this problem as it measures the water temperature where and, therefore, when it mixes the hot and cold water supply.

The problem gets even more interesting when various showers are connected to the same hot water supply, as the dynamics of the distribution network and the users are now interacting. When implementing feedback we need to respect the dynamics of the interacting systems.

1.4.2 Acoustic Feedback

The well-known phenomenon of acoustic feedback (see Fig. 1.8) can also be attributed to delay combined with feedback. A microphone placed in front of a speaker will cause an unpleasantly loud whistling sound, the particular pitch of which is related to the distance from the microphone to the speaker[5] and the amount of amplification that is placed in the sound feedback loop.

1.4.3 Boiler Water Level Control

A similar problem occurs when we try to automatically maintain the water level in a boiler using something like the toilet flush mechanism. In a boiler as steam is being formed, the water level will rise due to expansion but more importantly it will rise due to the formation of water vapor bubbles inside the water volume. Recall what happens when we put a pot with milk on the stove. As soon as it starts to boil, the liquid level increases rapidly although the amount of milk in the pot has not changed. In the boiler, as water evaporates, the total amount of water will begin to drop. When the perceived water level is too low, the feedback mechanism will add new, colder water. This cold water lowers the temperature of the water in the boiler, therefore reducing the amount of water vapor and hence shrinking the water volume. Surprisingly, this may result in a lowering of the perceived water level despite the supply of water. This is precisely the opposite effect of what was intended! A simple feedback based on the water level alone is not going to provide the right water level response; a more complex feedback solution must be found.

In most circumstances, feedback requires a very thoughtful approach. Eliminating feedback is certainly not the answer. In general, feedback requires systematic design in order to guarantee a successful outcome.

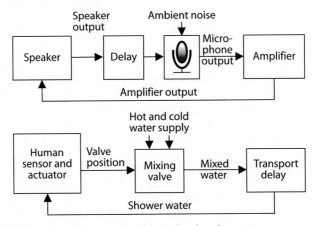

Fig. 1.8. A feedback loop in trouble: transport delay induced unpleasantness

---

[5] The distance determines both the delay, which is distance divided by the speed of sound, as well as the loss in sound power between speaker and microphone.

1.4.4 Birth and Death Processes

Consider a group of rabbits enjoying an unlimited food supply and no real predators. In these utopian circumstances, the birth rate (percentage of new individuals in the total population) will be larger than the death rate (percentage of individuals that die). As a consequence this rabbit population will continue to grow.

This case exemplifies a *positive* feedback loop which leads to unlimited and forever accelerating growth (an ugly prospect).

Assume now that the supply of food is constant, and that the birth rate is proportional to the amount of food available to any individual. In this case as the population increases the amount of food per animal decreases and so does the birth rate. At some instant in time the birth rate will equal the death rate. At that moment equilibrium is reached and the total population remains constant.

In this case, *negative* feedback is at work: an increase in the population (because the birthrate is larger than the death rate) leads to a decrease in food availability, and hence a decrease in the birthrate. Also, with a similar argument, a decrease in the population will lead to an increase in the birthrate. We say that the equilibrium (birthrate equals death rate) is *stable*, as a small deviation away from the equilibrium leads to a response in restoring the equilibrium.

Often, negative feedback is associated with stable equilibria and positive feedback is associated with instability. Though in general, this is a too simplistic point of view.

1.4.5 Manufacturing and Robots

The industrial revolution and its associated mass manufacturing of goods was enabled by the steam engine. The safe operation of the latter is due to a simple feedback device, the governor that regulated the speed. Unfortunately, the governor did not always work well. In some situations, large speed oscillations occurred, that were very detrimental to the downstream process the steam engine was driving. It appeared that this was due to how the engine was interacting with the production line. There was no simple solution. This problem lead to the first mathematical exploration of feedback in the mid 19th century. The economic importance of the problem was great and it attracted the attention of such giants in the world of science and engineering as Maxwell (see Chap. 8 for a more in-depth discussion).

Since the early days of the industrial revolution, the continued mechanization of manufacturing has been a major driver in the development of feedback. Today, manufacturing makes extensive use of robots, complex machines that execute repetitive tasks in a flexible manufacturing cell. New challenges for feedback design appear in the coordination of many robots. Also, the complexity of the robot motion in three dimensional space, coupled with the relentless drive for higher throughput speed combined with a need for high dimensional precision in the final product make even the heaviest steel robots appear like vibrating flexible structures. The combination of strength, speed, precision and flexibility in complex motion make for challenging trade-offs requiring non-trivial feedback design. We have the utmost admiration for how well elite gymnasts master this.

1.4.6 Feedback Design and Synthesis

The examples may already provide the hint that despite the fact that systems naturally contain feedback that it is not trivial to implement or engineer feedback properly.

Indeed feedback must be planned, and it must serve a particular purpose or objective, feedback for the sake of feedback is meaningless. Information on which to base the feedback (remember feedback reacts on the fact that its purpose is not satisfied) must be made available. This typically requires the provision of particular sensors that measure whether or not the feedback goal has been reached. The information from the sensors must be interpreted, and it must be clear how to impact the system under feedback so as to solicit a response towards satisfying the desired objective. This requires a means of translating feedback information into action, through devices or subsystems called *actuators*. All of these aspects do require design:

- What is the purpose of feedback, which objective has to be realized? Is it well-posed? For example, a truck cannot drive like a Formula-1 racing car, nor can a Formula-1 car be used to ferry a few tons of goods. Objectives must be matched with the available resources.
- Can the objective be verified? Which sensors should be used? There is no point in having an objective that nobody can measure or verify.
- How can sensor information be converted into action? Even if information is available, it must be sufficient to be able to decide on the next action, e.g. there is absolutely no point in telling a student that the essay was bad.
- How should one act on the system? Which actuators are necessary? Is there enough capacity (power, energy, force, …) to react adequately? Knowing what to do is good, but can it be done?

We will deal with all of these questions, and more, in the sequel.

1.5 Systems, Causality, Stationarity and Linearity

Before attempting the synthesis of feedback loops, it pays to understand system behavior, or to do some analysis of the dynamics of systems and feedback loops in particular.

System analysis goes via the signals that are associated with the system of interest. Signals and systems always go hand in hand; signals are derived from observing systems through the use of (measurement) systems and systems are analyzed through the signals they produce, see Fig. 1.9. If we are interested in signals themselves and more importantly the information they carry, we may not care to analyze how these signals are produced, as shown in Fig. 1.9a. For instance, if one is interested in the tide, there is a lot one can do without asking the question what is the system that generates the tide. One could simply study the motion over time, understand the maximum/minimum value, the frequency, the height, the tidal power and so on. One could even harness the tidal power and design a power converter, without immediately linking the tide to the position of Earth relative to the Moon and the Sun.

If we are mainly interested in the systems that underpin the signals, we need to look at the structure of the systems and how they interact, as shown in Fig. 1.9b. If you try

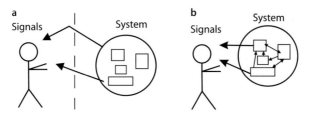

Fig. 1.9. Signals and systems

to build an irrigation system what matters most is the infrastructure, the components, their interconnection, their dimensions, the maximum forces they must withstand, not necessarily the precise dynamical behavior of water in the channels. In general, though both signal and system aspects are to be considered together.

Behavior

We use the term *behavior*[6] for the collection of all possible signals that *may* be observed from or associated with a system. Not all signals must be measured, even inputs (signals that systems act on) and outputs (signals that are/may be observed) are not always measured.

Not all the signals in a process are necessarily used in a block diagram, or system description. For example we may content ourselves with a mechanical description of the tide, using water flows and water levels, including the effect of the gravitational forces, but we may completely ignore the salinity, pH, temperature and color of the water. We determine the boundary of the block diagram, what is in, and what is out (hence considered as a part of the environment).

In our discussion of the feedback in the shower, the temperature of the water in the shower is clearly an output of the system. It is readily measured or observed, and it obviously depends on the temperatures of the hot and cold water supply as well as on the position of the mixing valve. The temperature of the hot water supply is an input, as clearly the temperature of the supply water is not influenced by the shower. The temperature of the water immediately after the mixing valve(s) is an output, but we cannot observe this temperature in general. Similarly the water temperatures of the hot and cold water supply are signals but they are not readily observed. The (accessible) behavior of our shower may simply be the collection of all histories of tap positions combined with the corresponding histories of shower temperatures.

So far in this shower example we only focused on water temperature. A more complete description must take water flow into account as well. In order to do so we must also consider the water pressures. For example, we may include a water pressure for the hot water supply and the cold water supply, and a water pressure in the mixing chamber. These signals could also be included in the behavior of the shower, or we could say that these are imposed from the environment.

---

[6] The term was introduced by Jan C. Willems, a thought provoking and inspiring teacher.

We could try to describe a system through its behavior. This is painfully impossible, as typically the behavior contains infinitely many signals, and requires observations over arbitrarily long periods of time[7]. Just imagine what it would take to observe all possible signals (remember these are functions of time) associated with a shower, not to mention trying to collect the behavior of an ecosystem. Though we cannot observe the totality of a behavior, nevertheless partial observations of the behavior is all we are ever going to get from a system. As a consequence, we will happily make hypotheses and provide descriptors for the total behavior based on our partial observations, that is after all the scientific method.

Despite their impossible generality, behaviors are important tools to understand system properties such as *causality*, *stationarity* and *linearity*. They provide us with the best opportunity to lay an axiomatic basis to systems theory.

Causality

A system is *causal* when for any signal in the behavior *the future cannot influence the past of the signal*. Because in our discussions signals are functions of time, causality is an all too natural a property. It is indeed not conceivable that the presently experienced shower temperature could possibly depend on future positions of the water taps. Indeed, our shower is a causal system.

Stationarity

A system is said to be stationary if for any pair of input/output signals that belongs to its behavior, a time shifted version of these signals also belongs to the behavior. For our shower (system) this more or less translates to saying that the experience of taking a shower does not depend on what day of the week that shower was taken (we are talking about water tap positions and water temperatures).

Linearity

A system or behavior is *linear* if the linear combination of signals in the behavior is also in the behavior. More elaborately, a linear combination of two input signals links with an output signal that is obtained by the same linear combination of the output signals associated with the specific inputs under consideration.

A linear combination of signals consists in operations that involve scaling (as in multiplying with a constant) and adding signals.

Clearly the shower system is not linear. Indeed we cannot even contemplate a linear combination of signals as the tap positions are obviously limited to lie somewhere between fully closed and fully open. So arbitrarily scaling tap positions is out of the question. Moreover for most taps, the valve characteristics are notoriously

---

[7] There is a notable exception, the class of systems described by so-called *automata*. At their core digital computers are automata, for which there are only finitely many possibilities and the behavior is hence countable.

nonlinear, the effect of opening a tap a little depends very much on how much flow is going through the opening already. Nevertheless, for any given position of taps, with a corresponding shower temperature, small changes in the tap positions will result in small changes in the shower temperature. These small changes will typically follow a linear relationship.

These observations hold quite generally. Most systems are indeed not linear, but for minor variations around some particular signal in the behavior, linearity will hold true.

Despite the fact that most systems are nonlinear, much of systems, feedback and control theory is nevertheless concerned with linear systems. Indeed most design and synthesis is performed using linear systems. Linearity enables efficient computation, and even nonlinear systems are typically approached from a computational point of view as a set of many linear systems. Linearity is an extremely simplifying property. It makes analysis and (feedback) synthesis so much easier. It is therefore invariably the place to start. Moreover, as will become clearer, often feedback provides a means of ensuring that linearity holds as a good approximation.

The System State

A very special class of signals, is the so-called *state* of a system. A state is a collection of signals, such that knowledge of the present value of this state combined with the present and future of the input signals provides us with enough information to be able to specify what the future is going to be. In other words, a state summarizes the relevant past.

For example in the toilet example, the water level in the cistern is a state signal. Indeed with this information and the future valve positions we can predict the future evolution of the water level.

A state (there is no *the* state for a system) is often not easily accessible, or directly available for measurement. It is however a most convenient vehicle to model and analyze a system with. Indeed, any simulation or computation of system behavior requires the construction of a state, as it amounts to having enough information for enabling a simulation.

Full knowledge of a state for a system is rather special, a rare event really, and because it contains all the information we need to know about the system (apart from the present and future input signals) to predict the future it provides us with enormous opportunity for control and feedback.

In the context of systems that describe the physical world, the concept of state can be most readily associated to a system's reservoirs for energy or materials. More abstractly, even if we do not have a physical concept of energy to relate to, the history of a system's response can function as a state.

These concepts will be developed further in Chap. 5.

1.6 Models

In general, obtaining a behavior for a system by collecting all possible input/output pairs of signals compatible with the system is an impossible task. We have to find a more compact way, something that can be communicated more easily. Nowadays we typically attempt to describe the behavior, and hence the system, through means of a computer program. In principle the algorithm could express a rule that allows us to decide when a particular signal belongs to the behavior or not. Even that is hard, and models often only provide a recipe to describe a behavior that supposedly represents the physical behavior of interest. The task of obtaining such a rule or algorithm is called *modeling*.

When discussing how to obtain equations or models for a system, it helps to focus on the process *from data to model*, that essentially underpins this. It is a manifestation of the scientific process: data is gathered and a model is nothing but a hypothesis consistent with these data. The unattainable collection of all possible data that can be derived from a system is its behavior. Our observations are by necessity a sub-collection from the behavior. In this rather simple and very general set theoretic framework, the basic ideas about modeling and the hierarchy of models and systems are reasonably clear. We will not explore this in depth, but it helps to keep in focus a number of concepts.

The rules are in general called *models*. A model is more powerful by the amount of input/output signals it is able to exclude and provides a better description of the behavior. Often we have to settle for approximate models, models that neither capture all the possible signals, nor exclude all those that should have been excluded, or that describe them only in some approximate manner, say to within measurement error.

Models do not have to be presented as computer programs. Programs are only the present convenient technology for unambiguous communication. For example the famous model by Kepler[8] for the behavior of planets in our solar system consists of three sentences[9]:

1. The orbit of a planet is an ellipse with the Sun in a focus of this ellipse.
2. A line segment joining the planet to the Sun sweeps out equal areas in equal times.
3. The square of the period of revolution around the orbit is proportional to the cube of the length of the minor axis of the ellipse.

This powerful model, incorrect as it may be, allowed Newton later to formulate the laws of gravity. From a control point of view, this is a rather boring model, as there is absolutely nothing we can do about it: it has no inputs. Of course, it embodies important knowledge if you want to launch some object into our solar system, or want to understand the tide.

As another example of a model, Newton[10] formulated his Laws of Motion of rigid bodies using three sentences. From Newton's laws of motion we can derive Kepler's model, and hence Newton's laws form a more powerful model for the motion of plan-

---

[8] Johannes Kepler, 1571–1630, German mathematician and astronomer. He studied at the University of Tübingen and lectured at the University of Linz.
[9] This example is adapted from J. C. Willems's work (1997).
[10] Isaac Newton, 1643–1727, English physicist and mathematician, one of the greatest minds of all time.

ets. Even from a control point of view, there is absolutely nothing boring about Newton's laws of motion.

1.6.1 Modeling

The most common way to deal with signals and systems is to represent them by models, and our models are invariably biased by our interests, or the actual purpose we have for the model. The latter is always the place to start, without a clear purpose any modeling exercise is doomed to fail (because you cannot tell if you have succeeded or not).

Modeling requires consideration of aspects such as:

- What is the model for? What questions need answers?
- What physical principles are at work?
- Is there already a computer algorithm that must be verified? What can be challenged? Is there a preliminary model that needs refining? Is there a very complex model that needs to be simplified?
- Is there a conceptual block diagram? What variables matter? Which are less important? Can the block diagram be challenged?
- What experiments are required to identify missing information? What experiments are feasible?
- What signals are needed? What can we expect (range, repeatability, stationarity, randomness)?

Once the signals of interest are identified, their measurement process requires attention, and its purpose is subserviently linked to the ultimate purpose of obtaining the model. In particular, the following issues should be taken into account:

- What should be measured? Is a direct measurement feasible, or is the signal of interest to be inferred from measurements? How reliable is this inference?
- How well should various signals be measured?
- What sensors are required? Determine the range, and accuracy and the response speed of the sensors. Is redundancy required?
- How reproducible are the measurements? Are all the relevant aspects captured?
- Are the measurements dependent? Is there redundancy? Can the measurement process be calibrated independent from the modeling task?

Despite the fact that modeling is as old as our desire to understand and rule the world, modeling is still as much an art as a science, and experience plays an important role.

Perhaps surprising, perhaps not, many signals (even from totally different physical domains) and the behaviors of many systems are captured by the same or very similar (computer) models. Of course, the physical reality, the variables, the parameters, the units and scales may all be different; but the essence of the behavior is not. The fact that there is such analogy between different systems, provides a great opportunity for abstraction, and allows us to build a unifying framework for analysis and synthesis of systems.

1.6.2 System Interconnection

Systems typically comprise subsystems interconnected with each other (consider the block diagrams!). To build a new system we may interconnect subsystems so that some of the outputs of one subsystem become inputs to another subsystem, more generally interconnection of systems simply means sharing of signals. A very simple interconnection of two systems is to have the output of one system be the input to another system. This is called a cascade or series connection. It is more interesting to create (feedback) loops. In the case of feedback, some of the outputs derived from a subsystem will influence (through the feedback loop) some of its inputs, and hence the outputs.

It is intuitively clear that an interconnection of subsystems leads to a new behavior that is constrained with respect to the collective behavior of the subsystems. By necessity the interconnection imposes restrictions on some signals, where there were no such restrictions before. Inputs are free signals, from the point of view of the system to which they are inputs, but by virtue of an interconnection, some inputs are now outputs from another part of the system, or shared, and hence no longer free as far as the interconnected system goes. So we can say that the collection of signals that are supported by a feedback loop is less rich than without the feedback. Indeed a successful feedback loop eliminates unwanted signals, and emphasizes the preferred behavior. In contrast to this simplification, from an analysis point of view, feedback typically complicates matters, but that is no excuse for not using it. Without feedback, more signals are free, hence easier to understand. Of course, feedback also leads to more interesting behavior and sometimes unexpected behavior, and that is precisely the *joy of feedback*.

1.7 The Basic Control Loop

The basic control loop consists of sensors to measure what matters and a capacity to interpret the data and decide on an action, that is then implemented using some actuator. It mimics very much the way humans or animals control tasks that they have intended to complete: sense, interpret, act. Consider for a moment, by way of example, the relatively simple control task you are performing right now, namely reading this text.

- The initiation for this activity, that is to have the intention, make the time and decide to read this text have already been performed.
- Cognitive processes allow you to read the text, see the words and put them in context;
- Memory is activated to save some information, but more importantly to recall other information from your experience as triggered by the new information you are acquiring;
- Colored by emotion, intelligence allows you to interpret the reading based on your previous knowledge and organize the new information, and store it for further use;
- Motivation urges you to continue reading, and to concentrate (blocking other sensory inputs) and to persevere (well we hope so) and last but not least
- Your eyes, fingers and posture are coordinated to enable the reading at the desired speed.

Not surprisingly, engineered control systems at least conceptually mimic much of this structure and organization. Indeed, part of the study of control engineering is motivated by pure curiosity to better understand the marvels of human control and communication. In the 1950s, Norbert Wiener[11] proposed a nice, but no longer popular word for this *Cybernetics*[12]. The dream of creating fully autonomous machines is still a dream. Nevertheless, in some areas, control has been instrumental to deliver autonomy of task execution, with incredible accuracy, reliability and repeatability.

The basic control (sub)system is composed of (see Fig. 1.10):

- *Sensors and data acquisition systems* (DAS). Sensors respond to physical stimuli (heat, light, sound, pressure, magnetism, motion and so on) and record or transmit a response. Typically they include *transducers*, that convert the sensed signal into another signal domain, nowadays often in a digital format.
- *Digital signal processors* (DSP), perform scaling, filtering smoothing or any other preprocessing function. They could be a stand alone component in the communication chain between the data acquisition system and the control computer, or be incorporated as part of the latter, or as part of the former.
- *Controller*, a dedicated microprocessor or general purpose computer, with memory to store the procedures, algorithms as well as data, equipped with communication links for supervision, and/or (re)programming as well as receiving information from the DAS and sending information to actuators. The control computer computes and supervises the control actions. The algorithms either directly use or are based on knowledge of the process model (the object under control), the signals, as well as the objective of the control task at hand.
- *Actuator*, which receives the commands from the control computer, and transform these, typically using additional power sources, into inputs that drive the process.

The brain of the control subsystem is the control computer, but the coordination with sensors and actuators, based on a thorough understanding of the process under control, is fundamental for the controlled system to perform as desired.

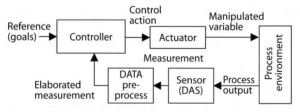

Fig. 1.10. Components of a basic control loop

---

[11] Norbert Wiener, 1894–1964, U.S. mathematician and engineer. He graduated from university with a degree in mathematics at the age of 14. Obtained his PhD at the age of 18 from Harvard University. He became a professor at MIT and laid the foundations for much of modern signal processing and control theory. He is the father of Cybernetics.

[12] Cybernetics is a Greek word from "kybernetes" (from kybernan to steer, govern+English -ics) meaning "steering a ship": the science of communication and control theory that is concerned especially with the comparative study of automatic control systems (as the nervous system and brain and mechanical-electrical communication systems).

Control systems vary hugely in complexity. In primitive control systems the control computer is nothing more than a connection between sensor and actuator, like the float in the toilet flush mechanism, in other environments like controlling the world economy it is the human-in-the-loop that decides on what is the next action. Nowadays it is quite common that sensors, actuators and control computers are not co-located, but communicate through a communication network or communication (sub)system just like in the human control loop, the sensors, muscles and brain are connected through the nervous system.

In some cases, the control can be performed based on the knowledge of the process and the required goals, without using information from its actual evolution. In this case, the control is performed in open-loop, hence without feedback.

1.8 Control Design

Synthesis of a control task always starts with the control objective. Is automation desired, necessary, beneficial, essential? What purpose must be achieved? What are the alternatives?

For example, we want a computer assisted brake in a car to achieve better safety, to achieve less trauma on the roads.

A brake is of course an essential safety device in the car, so reliability will be of the utmost importance. There is a lot of experience with sensing and actuation of a brake, so we may decide to re-use these components and concentrate on how to achieve better braking, with the main objective to deliver maximal deceleration regardless of road condition. In general, even when the purpose is very clear, questions such as what to sense, how to sense, and what to actuate and how to actuate need to be addressed. There is a (nice positive) feedback loop at work here: new developments in sensing and actuator technology drive the development of automation, and higher expectations in automation drive the development of sensing and actuator technology.

Modern control design is typically based on a model of the process under control, so that we can test and evaluate the performance of the system without having to experimentally verify each option under consideration, which would be prohibitive. There is a non-trivial interaction between modeling and control. The better the model, the more we can try to do with control. The better the control, the less we need to know about the model outside the controlled behavior. Some aspects of the model are irrelevant for control, others are critical. In control applications, overkill on the modeling side is quite common, as the inclination is to model for everything, not just for control decisions.

Braking in particular is a difficult process, how does a rubber tire interact with another surface (not only a concrete road, but a thin layer of ice, a water logged surface, a mud trail, gravel)? Should we sense the type of road, or can we get enough information from the forces exerted on the wheel? How important is it to know the pressure inside the tires? How are the brakes on the different wheels coordinated? The weight distribution in the car is going to influence the brake capacity on each wheel. Modeling is non-trivial, but rather essential in this case and trial and error is not going to get us very far.

Given an acceptable model for the process including models for the sensors and actuators and communication channels, and equipped with a clearly defined objective, the actual design of a control algorithm can start. The tools vary from heuristics based on experience, to hybrid optimization techniques. The advance of theory in control is such that a great deal about what can and cannot be achieved, i.e. the fundamental limits of system interactions are now well-established. Often the design process will be iterative. Solutions are proposed, tested against the design models, then against more elaborate models, and the objectives refined the process iterated until a satisfactory design is achieved.

Once the control process has been settled, it can be evaluated and commissioned in the physical world. Prototypes are built and tested (well beyond the normal operating conditions), and failure modes identified. On the basis of these outcomes the design is either refined or passed onto production and its final realization.

1.9 Concluding Remarks

Systems theory is the study of signals and systems and their models: how to describe them, analyze and classify them. At the heart of control is the power to build new systems, with new behavior by *simply* interconnecting systems. How do systems depend on each other, how do they interact? From an engineering point of view, most important is how to synthesize or design new signals and systems with desirable properties, how to use systems and signals as bricks and mortar.

Most of this study is carried out through a high level of abstraction, the use of mathematics, with computation at the heart. The object is to study the intrinsic properties of signals, systems or models, not their particular physical realization.

A large body of work in the general study of systems theory is devoted to the process of extracting from observations a model that describes these observations, the so-called *data to model* transformation, or modeling. In systems theory we are not really concerned with the particulars of the implementation of the process from data to model; e.g. how data are collected, or stored whether it be in the brain, or in a computer or with pen and paper. Rather we are interested in the intrinsic, that is to say independent of representation, issues regarding data, the systems they are derived from and the models they support. Objects must be studied such as the relationships between data, or that may exist in data sets, relationships between models, between data properties and model properties, between purpose of model and required data.

This study of signals, systems and models is not really new. Since the beginning of civilization we have been interested in explaining what we observe, that is identifying useful relationships between observations, which is nothing but the construction of (different) models (in our mind) for the observed world. This is the essence of the ongoing scientific process. Not only does the power to explain appear to endue us with authority and a strange sense of achievement, as if we had created the world around us, but more importantly our explanations are used to change how we interact with and indeed change our physical environment. What is perhaps more recent, and particular to systems theory, is that we study signals and systems in a much more abstract way, not referring to their physical realization, but rather through their general properties and structures revealed through mathematical analysis.

Significant achievements were driven by speed control in engines, telecommunications, auto-pilots, Moon and Mars missions, the complexity of large scale chemical processes and the large scale utility networks. More recently theory and practice are spurred on by the potential of the internet and wireless sensor-actuator networks for use in home automation, human body networks and intelligent transport or smart infrastructure systems.

Entire books, from many different angles, econometrics, statistics, psychology or engineering are written on the topics that we touched upon. After all signals are fundamental to observation. Moreover modeling is innate to human existence: *cogito ergo sum*[13], it is one of the main pillars of the scientific process, fundamental to all of science and engineering.

1.10 Comments and Further Reading

Systems and control theory is a relatively young discipline, that finds its origins in cybernetics, the latter is a term coined by Norbert Wiener (1948, 1961) just after World War II. Systems theory shares some common aspects with dynamical systems theory, a branch of mathematics, which in some sense is an abstraction from thermodynamics. The distinction between systems theory and dynamical systems theory is that in systems theory we always deal with inputs and outputs or so-called *open* systems, whereas in dynamical systems theory inputs do not play an important role. Inputs and outputs are critically important when considering systems as building blocks to construct larger systems through interconnections.

The notion of system was possibly first coined within the context of thermodynamics. It is definitely an overused if not abused term since it is used in most sciences.

The study of electrical networks, or electrical systems is normally credited with the origin of the systematic study of systems in a control theoretical sense. It is in electrical systems, in either power or telecommunications applications, that engineered systems first reached a level of complexity that demanded a more abstract treatment of system synthesis. (See e.g. Anderson and Vongpanitlerd 2006; Belevitch 1968.)

Control engineering has always been technologically driven. At the dawn of the industrial revolution the (incomprehensible) behavior of the governor of Watt's steam engine provided the first stimulus for system analysis. At the end of the 19th century and the start of the 20th century the advent of power and telecommunication networks provided ample motivation. In the 20th century, avionics, and the space race in particular stimulated much development. Today the field is being spurred on by wireless sensor/actuator network technology and the potential of a new bio-engineering industrial revolution. A history of the field, at least up to 1955 can be found in Bennet (1979, 1993).

A course on signals and systems is nowadays a mandatory component in most electrical engineering curricula, being also compulsory under similar names in many other engineering curricula, like aeronautics, mechanical, chemical or telecommunication. Illustrative examples and numerous versions of open source courseware can be found

---

[13] I think, hence I exist; famous expression from René Descartes, French philosopher and mathematician, 1596–1650.

on the world wide web. Text books are numerous, for example Haykin and Van Veen (2002) and Oppenheim et al. (1982). Most require a working knowledge of elementary linear algebra and calculus. A more modern approach, perhaps somewhat more broadly accessible, can be found in Lee and Varaiya (2003).

Ideas to generalize systems theory such that it seamlessly can deal with biology, physics, thermodynamics and so on were initiated by von Bertalanffy (1980).

The development of systems theory starting from the behavior of a system is relatively recent development, and is due to Jan C. Willems. For an exposition of these ideas see for example Polderman and Willems (1998). The so-called *behavior* approach is elegant, and provides an accessible and modern axiomatic basis for the study of systems.

There is a very active research community in the field of systems and control. The Institute of Electrical and Electronics Engineers (IEEE), *Control Systems Society* (CSS), has provided a focus for the control systems community since 1954. The *International Federation of Automatic Control* (IFAC) is an international organization (see www.ifac-control.org) dedicated to systems theory and its applications since 1957 in the broadest possible sense. The IFAC runs many symposia in the area and a world congress every three years, the latter attracts well over 2 000 researchers. The *Mathematical Theory of Networks and Systems* (MTNS) is a biennial conference that places the mathematical aspects of systems theory and interconnection through networking on center stage.

There are a number of large multinational companies that make control and automation their primary business and service. Companies like IBM, Honeywell, Rockwell Automation, Omron and Siemens are all major players. Software platforms like the commercial Matlab^{TM} [14] and LabView^{TM} [15] as well as the open source SciLab[16] support teaching, research, and engineering development in control, automation and systems engineering more broadly.

---

[14]Software from the MathWorks company *http://www.mathworks.com/*.
[15]Software obtainable from National Instruments *http://www.ni.com/labview/*.
[16]*http://www.scilab.org/*.

Chapter 2: Analogies

- 2.1 Introduction and Motivation
- 2.2 Signals and Their Graphical Representation
- 2.3 Signals and Analogies
- 2.4 Example Systems
- 2.5 How Similar Is the Behavior of These Example Systems?
- 2.6 Discrete Time or Continuous Time?
- 2.7 Analogous Systems
- 2.8 Combining Systems and Splitting Distributed Systems
- 2.9 Oscillations
- 2.10 Comments and Further Reading

Chapter 2

Analogies

> *The road is full of vehicles moving along it.*
> *The river transports water, with leaves, small insects, fishes and so on.*
> *You cannot see it but information is flowing through the telephone lines.*
> *Your blood is flowing through your veins allowing you to read.*

2.1   Introduction and Motivation

Spontaneously we compare new things with what we have observed before, with the things we are already familiar with. This is a natural form of abstraction. By recognizing similarities or analogies in the way different phenomena happen we start to impose some structure on the world around us.

In this chapter we describe a number of simple processes that at first sight appear to be very different in nature. At least they will deal with physically different things; the warming of an oven, the flow of water into a leaky tank, the charging of a capacitor and a particular computer algorithm. Yet when we observe how they behave over a period of time, when we measure the temperature in the oven, the level of water in the tank, the amount of electrical charge in the capacitor and the numbers generated by the computer algorithm then, at least at the level of these observations, they will appear to behave very similarly. The analogy goes far deeper than simply these measurements, to the point that we may substitute one physical system for another without much loss in understanding what actually goes on. We say that one system *models* the other. Because the computer algorithm is by far the simplest system to be implemented in such a manner that we can communicate experiments and results verifiably and reliably, we like seeing it as the *model* of choice for the other physical systems, whether it be the oven, the water tank or the capacitor.

The model itself is not so important. What is important is that a model provides a means to describe a partial behavior of a physical system in a precise and concise manner (and more concise than the collection of observations themselves).

Models are to systems engineers, what plans are to architects. A plan describes a building, bridge or an airport using pictures that have been drafted according to agreed standards. The plan is a concise and precise way of communication. It conveys what has to be constructed. In a similar fashion models describe in a concise and agreed manner the behavior over time of a physical system. Architects may also create a *model to scale*, using say paper or plastic, or a three-dimensional computer graphics rendering of the building. Similarly systems engineers may reproduce behaviour using a mechanical and/or electrical analogue that emulates the behavior of the actual process. Nowadays however, models run on digital computers that execute computer algorithms embodying the mathematical models. These are what we will refer to as *computer models*.

Observations can be both of a qualitative nature and a quantitative nature. For instance, referring to the ambient temperature we can qualitatively say that it is cold, warm or hot, or, more precisely, we can say it is 21.5 °C. We are more interested in quantitative information, and we will often resort to quantitative characterizations of qualitative properties (because our models live in computers). Quantitative information and computer models play an ever increasing role in science and engineering, because computers allow us to compute things fast (and reliably) and perhaps most importantly computer models can be easily communicated.

In order to be able to compare measurements we will first discuss some conventions on how to represent them as pictures, as this communicates more easily than sequences of numbers and also more easily than computer programs.

The rest of this chapter is organized as follows. First we will talk a little about conventions, to make sure we understand signals and their graphical representation, as we will make extensive use of such pictures in the book. Then we will describe the examples, the oven, water tank, capacitor charger and the computer algorithm. We will compare them, and resolve to what extent they model each other. Finally we will discuss how we can use these simple systems as building blocks to construct and/or model more complex systems. In the further reading and discussion section we will provide pointers to the extensive literature about understanding the true nature of analogies, and the literature on formalisms that capture or exploit these analogies to enable computer modeling of physical systems.

2.2 Signals and Their Graphical Representation

A signal is a record over time of a particular variable; a temperature in a room, a water level along a river, the amount of charge in a capacitor. At each instant of time, there is a value for the variable of interest. Each observation or data point gets a unique identifier, *an index* with which it can be retrieved at a later stage (the time instants). Typically there are many ways in which to index a particular set of data, depending on the way we view the data. Sometimes an index comes almost automatic. In mathematical language signals are *functions* that associate to each index value from an index set a unique value from a set of possible values. If the index is or contains time[1] we call it a signal. In case the index set considers time in distinct values, i.e. as discrete, the signal is called a (time)series.

Take for example a mercury thermometer, as shown in Fig. 2.1a. The mercury level varies with time. Its value can be represented as in Fig. 2.1b, where the horizontal arrow indicates increasing time and the vertical arrow indicates increasing values of the measured temperatures. These measurements may be recorded on a magnetic tape, (Fig. 2.1c). In this case, the signal is an analogue continuous time signal: both time and value are considered over a continuum.

Assume now that we only consider the temperature from time to time. Anytime we look at the thermometer a temperature value is read. This data can be represented as a table with two columns (see table in Fig. 2.1e). We can place the index or the time at which the temperature is read in the first column, and the observed temperature at the corresponding time instant in the second column. This is not very convenient. Tables are rather boring to look at, and do not really convey the variability well. However,

---

[1] There may be more in the index than time, for example the temperature in a room depends on space and time. We consider the collective of the temperature over space as the value of the signal.

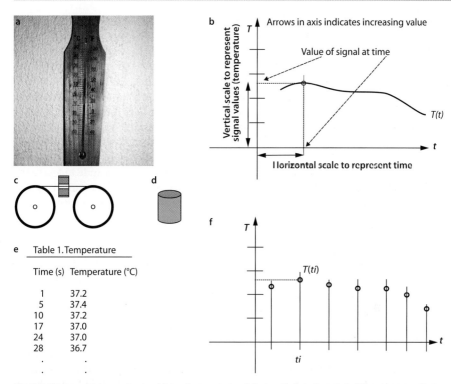

Fig. 2.1. Temperatures as observed by a thermometer (**a**) recorded and stored either as an analogue (**b–c**) or discrete (**d–e–f**) signal

tables do clearly indicate the finite precision with which the temperature is recorded, e.g. because of the finite number of digits used, like 37.2 °C.

Tables are a bit like a digital watch. A digital number representing time does not nearly convey the progression of time as well as the motion of the hands on a clock. Those tables are represented as pictures in this book. The first column, time, will be the length along the horizontal row according to some scale. The second column, the measured variable, will be represented by the length in the vertical direction. So with each row in the table one point in the picture corresponds to the coordinates time and value expressed as lengths in the horizontal and vertical direction on the paper respectively. This idea is graphically represented in Fig. 2.1f. All signals in this book are really nothing more than tables, or collections of points on a plane. Nevertheless, even a collection of separate points does not really convey the idea of continuous change that we intuitively[2] associate with the variables of interest. The points are then often explicitly connected to form a line on the paper. Such pictures are often called *graphs*. Graphics make it easier to follow the evolution of a variable.

---

[2] This intuition may well be incorrect. The fact that our physical world is better described using discrete quanta for time, space and so on is precisely the objective of quantum physics.

Signals represented in tables are denoted as discrete time signals. If their value is also discretized they are denoted as digital signals. They can be easily recorded as strings of data in a digital recorder (Fig. 2.1d).

Sometimes we measure more than one variable at a time and so we can index the measurements with time, and an identifier for each variable. Even in those circumstances we often refer to the collection as a single signal and not as different signals. It is understood that the signal will have a set of different values (one for each component in the signal) associated with each time index. For example we may want to discuss the temperature and the atmospheric pressure over time. The signal has two values at each instant of time. If we consider temperature at many different locations across the globe, we can index with three scalars using a longitude and latitude to locate the position and a time. Again the signal value is the collection of all the values at a given time. Such signals are rather difficult to display, but that is exactly what weather map movies do.

> **Signal**
>
> A signal is a function from an index set usually representing time to a value set (say temperatures) that assigns a single element of the value set (the temperatures at that time) to each member of the index set (each time instant).

More on Different Signal Representations

A sheet of music, an ordered list of notes can be viewed as a signal. The index set is represented rather specially. First there is the order $\{1, 2, \ldots\}$, which indicates the sequence in which the notes have to be played and for each note there is a symbol indicating the length of the note in the sequence. Together they define the time index. The value set is the collection of tones or frequencies of the notes to be played. Interpreted by a musician on a musical instrument, the signal will (hopefully) become music to

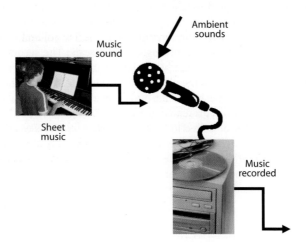

Fig. 2.2. Recording music

our ears. This music we hear is a signal: the air pressure on our ears over time. The sound can be recorded onto a compact disk as a string of bits that represent in digital format the air pressure (music) on a microphone at a sequence of given times, called *sampling times*, times at which the signal was measured or sampled.

Here we have three completely different representations of the same object of interest which is the music we can hear. It is represented as a sheet music, the air pressure on our ears over time, or a sequence of zeros and ones in a digital file or on a compact disk.

Both the sheet music and the compact disk recorded signals are *sampled data* signals. The index set is a set of discrete time values. Not only is the index set discrete, also the signal values are derived from a discrete set of values. In the sheet music only certain notes can be represented, as the composer follows an agreed convention. Similarly in the compact disk only an agreed range of integers is available to represent the measured air pressure, typically an integer between 0 and 4 095. This obviously involves an approximation, as there is no reason to believe that air pressure can only take on 4 096 ($= 2^{12}$) different values.

Using an older recording technique, the air pressure may be represented as an orientation of a magnetic particle on a magnetic tape. In this case, we have a signal which has the distance on the tape from the start of the tape (which becomes a time by playing the tape at a certain agreed velocity) as index set and the orientation of the north pole of a magnetic particle on the tape as value set. We say that the signal has a continuous index set. If the index set represents time, it is known as a *continuous time* signal. Since the orientation is an angle, the value set is also continuous. In this case we say that the signal is an *analogue signal* (see Fig. 2.2).

So far, we have considered signals denoted as:

- *discrete*, if its value set is discrete. It is called an *analogue* signal if its value set is a continuum.
- *continuous time signal*, when the index set represents time as a continuum and a *time series* or *discrete time signal* when the index set represents time as discrete values.
- *sampled data signal*, if it is taken from a continuous time signal at some given instants of time (regular or irregular); that is, its index set is discrete. A signal is often called *digital* if both its value and index sets are discrete.

Neither of the recorded signals, on tape or on the music CD, represents precisely what we would hear as music. At least the older tape recording technique retains the property that music appears to be a continuous signal over time. But the sheet music or the compact disk recording both appear to be rather qualitative different in that they have both a discrete index set and a discrete value set. They are sampled versions of the actual music. The signals are related to each other through a system comprising the musical instrument, the air and the sensors (our ears or a microphone) used to observe the music.

Why and how these signals are appropriate representations of the music forms part of the study of *signal processing*.

Notice also that the nature of the signal in the sheet music is very different from the nature of the signal on the compact disk. The sheet music represents first the order of the notes, then the duration of these notes (tones or frequencies) as well as the associated energy (leaving ample room for artistic interpretation). The corresponding air pressure as produced by the musical instrument is recorded onto the compact disk. The connection to frequency has seemingly disappeared. Signal processing can reveal the link and recovers readily from the recorded signal the spectral content, indicating that the notes are indeed not only of a certain frequency, but also revealing much other structure like timbre.

Finally this example serves to demonstrate that signals and systems are closely linked, inseparable really. The sheet of music does not become music unless it is played on a music instrument, and it cannot be recorded unless a microphone is available to measure it. The recorded music signal on the compact disk is as much a representation of the sheet music signal as it is of the system (musical instrument(s) and microphone) that produced and captured the music.

2.3 Signals and Analogies

When looking at a record of a signal represented as a graph, for instance Fig. 2.3a, it conveys an impression of a quickly changing variable that first trends up, then down and up again as we move forward in time, that is to the right in the picture. It is a somewhat rough looking graph. In Fig. 2.3b there is a similar graph, only a lot smoother. The graph below is just a collection of points, where the points are equally spaced in

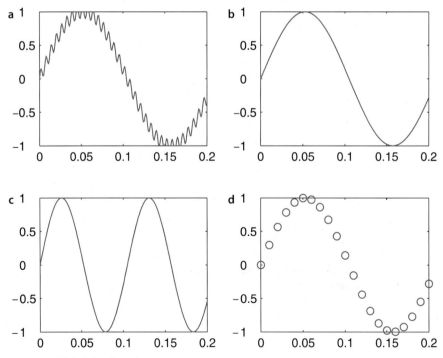

Fig. 2.3. Unlabeled signal graphs

the time direction. It appears to be a *sampled* version of the graph above it. The points (circles in the picture) are smoothly connected by the line in the graph above it. Finally the graph in Fig. 2.3c, oscillates between the same limits appearing smooth but moving up and down twice in the same interval.

These different graphs presented in Fig. 2.3 may actually convey the same information. Indeed, what is information in a signal and what is not, very much depends on who is observing the signal, and what the expectations are. It may indeed be the case that the roughness in the graph in Fig. 2.3a is appreciated as an artifact of the way the actual signal of interest, the one in Fig. 2.3b is actually measured. To another observer, both pictures may indeed convey a totally different impression since that observer may have no expectation as to the relationship between the two pictures.

One possible interpretation of these signals is that we are listening to a pure tone from a pitch fork, which produces the signal on the top right. If it is picked up by a microphone that also picked up some screechy noise in the recording environment, the combination produces the signal displayed on the left. The bottom left-hand picture is the same tone now reproduced from a tape recording that is played back at twice the speed of its recording speed. The graph on the bottom right is the same tone sampled at discrete instances of time or 20 equidistant discrete samples over the indicated time interval.

This effect of interpreting graphs or signals using preconceived expectations is really underpinning the idea of analogy or similarity. Often we see things in pictures or signals that are perhaps not really there, because our brain processes the data matching it to previously observed data. Let us look at the typical example[3] shown in Fig. 2.4. Depending on your frame of mind you will either perceive a young lady or an aged woman. The information is the same but the "brain processing" is different.

Fig. 2.4. A young lady, or is she old?

---

[3] This is a classical ambiguous picture, with a very long history, discussed in the work of Boring (1930), also associated with W. E. Hill's illustration "My Wife and My Mother-in-Law".

The nature of similarity is through observation and matching what has been observed with what has been experienced. This has the natural benefit of organizing the total information available, eliminating unnecessary clutter and reinforcing what has been learned already, but it has also a detrimental aspect as truly new information may be misclassified or totally ignored.

Analogies play an important role in the study of signals and by implication systems, as we experience systems through the signals we observe from them.

> **Signal Analogies**
>
> Two signals are analogous if their graphical representation is exactly the same after scaling the units (time and/or value unit).
>
> Two signals show some analogy if they present similar characteristics such as: size, duration, trend, envelope, average …

2.4 Example Systems

Here we describe in turn some simple dynamics (signal variations over time) as they occur in the heating of an oven, the filling of a leaky tank, the charging of a capacitor and finally a computer algorithm. In each instance, we only concentrate on a part of the overall system associated with a variable that is relatively easy to measure, respectively the temperature in the oven, the water level in the tank and the voltage level of the capacitor. In the computer algorithm we will be repeating some simple algebra and the signal will be a number indexed by how many times we have repeated the algebraic operation.

2.4.1 Heating an Oven

An oven (like a kitchen oven, see Fig. 2.5a) is a well-insulated chamber, so that heat[4] does not leak easily from it, and to which heat can be supplied from some energy source.

In order to better understand its behavior let us consider a physical schematic representation or model, as shown in Fig. 2.5b.

As the heat is supplied the temperature in the oven rises. Suppose for the moment that the oven is simply filled with air. The temperature of the air in the oven is an indicator for the amount of heat energy stored in the air inside the oven.[5] Despite the insulation, some heat does leak from the oven. The hotter the air in the oven, the more heat will leak from the hot oven to the colder environment. Eventually, the tempera-

---

[4] We use heat here as a term for thermal energy. Strictly speaking, in thermodynamics, heat is a term reserved to indicate the transfer of thermal energy between different objects.

[5] Another way of looking at temperature is to consider it as a measure for the kinetic energy of the air molecules, which is how *heat* is stored in the air. The hotter the air, the faster the molecules are moving and bouncing against each other and the walls.

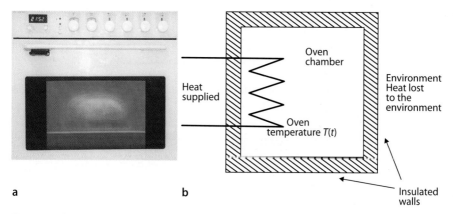

Fig. 2.5. A schematic representation of an oven

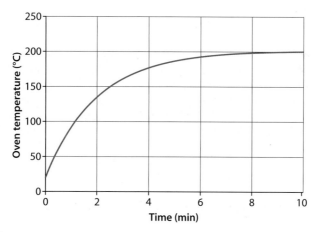

Fig. 2.6. Heating an oven; oven temperature against time

ture in the oven reaches a point where the amount of heat supplied equals the amount of heat that leaks from the oven. At such a point in time, the temperature of the air in the oven stabilizes. The first law of thermodynamics states that there is no loss of energy. Here, the amount of heat supplied equals the amount of heat stored (in the air) plus the amount of heat that leaks from the oven (and is stored in the walls and the environment).

If we now stop supplying heat to the oven, then the temperature in the oven will slowly decrease as the heat leaks from the oven. Eventually, the temperature in the oven will equalize with the temperature outside the oven. This is a manifestation of the thermodynamic *principle of equilibrium* (Gyftopolous and Beretta 1991) namely that objects in thermal contact eventually reach the same temperature.

When measuring the temperature in the oven during the heating cycle we would see a signal as represented in Fig. 2.6.

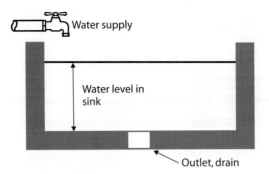

Fig. 2.7. Filling the sink with the drain unplugged

2.4.2 Filling the Kitchen Sink

Let us consider the filling of a bath or a kitchen sink from a tap without having closed off the drain completely, so that there is an outlet flow. The situation is schematically represented in Fig. 2.7. As long as the outlet flow is less than the inlet flow the water level will rise. The outlet flow depends on the geometry of the drain, but given it is fixed, the magnitude of the outflow is determined by the pressure of the water in the sink. The higher the water level is the higher the outflow. Assume that the sink is deep enough to allow the water level to rise so high that the outflow equals the inflow without causing an overflow. In this situation the water level will stabilize and become constant in the sink. The amount of water stored in the sink now remains constant, the inflow being equal to the outflow.

The experiment of filling the kitchen sink very much repeats the story about the temperature in the oven. In their respective contexts, the signals water level and temperature play exactly the same role. The water level captures the storage of water, and the leaking of the water through the drain. The temperature captures the storage of heat inside the oven, and the leaking of heat from the oven to the environment. Additionally, the actual measurements, see Fig. 2.6, for the oven temperature over time can look identical up to some scaling to the water level in the sink, i.e. with appropriate scales on the horizontal and the vertical axis Fig. 2.6 represents both the depth of water in the sink and the temperature of the oven. That some scaling is essential is obvious. As in one case we measure depth of water in the sink for which we use the meter for measuring and in the oven the temperature is measured in Kelvin or Celsius degrees.

2.4.3 Charging a Capacitor

Rechargeable batteries store electrical energy in the form of chemical energy, which can be used in turn to drive an electric circuit. Similarly electric charges can be stored in a capacitor using purely electric means. When charging a battery or a capacitor, the capacitor charger supplies the capacitor with electric charges from a constant voltage source (an electrical energy source, typically derived from the mains via some adaptor). The amount of charges in the battery or capacitor is indicated by the voltage across it. This voltage level is easily measured. Initially as the capacitor is discharged, that is its voltage level is well below that of the supply, the charging process is easy and electric charges flow quickly into the capacitor. This is because the flow of charges into the capacitor is proportional to the voltage difference between the supply and capacitor voltage. As the capacitor charges, its voltage increases and as a consequence it gets harder and harder for new electric charges to flow into the capacitor. When the capacitor's voltage level reaches that of the supply no further charge transport can occur, and the voltage level of the capacitor settles at that level. The voltage signal can again be represented by Fig. 2.6. Again the axes will have to be rescaled and relabeled, but that is all.

The signal in this example, the voltage, is also a measurement for the amount of energy stored in the capacitor. Unlike the oven or the sink examples, here there is no leaking of energy out of the capacitor. Indeed, the nature of the equilibrium reached is more akin to what happens when the oven is allowed to cool down and comes in equilibrium with its environment. Here the supply of electrical charges is proportional to the difference of the voltage levels between the supply and the capacitor.

2.4.4 Computer Algorithm

Let us consider the following simple recipe, or algorithm:

1. Pick any number, call it y.
2. Replace y by the new number $0.8\,y + 0.2$.
3. Repeat.

The first few iterations of this algorithm, starting from the initial $y = 0.2$ and $y = 3$, are represented in Fig. 2.8. The time is here the number of times the algorithm has been applied. The value of y at each iteration is a signal. Apart from the fact it is a discrete sequence of numbers, its appearance is very much like that of Fig. 2.6.

It appears that the algorithm reaches an equilibrium or a steady state, regardless of which initial number it started from. Indeed, if we repeat the algorithm, then with every new iteration the number gets closer to 1, the difference decreases monotonically. Also, it is not hard to see that if we started with $y = 1$, then the signal remains constant as indeed $0.8 \cdot 1 + 0.2 = 1$. These two properties define what we mean by a (constant) steady state or equilibrium.

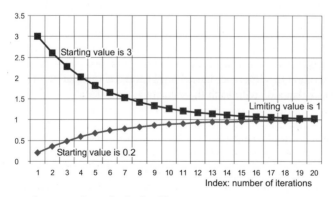

Fig. 2.8. Outcomes from repeating a simple algorithm

The algorithm's similarity with the oven, or water level is striking. Each time the recipe is applied, the present number is reduced proportionally by a factor 0.8 (it leaks away) and is replenished additively by a constant +0.2 (energy or water supply). In the end, the algorithm reaches an equilibrium in which the effect of leakage is negated by the supply. This is clear, at the equilibrium, the number 1 leaks to 0.8 and is replenished by 0.2 to become 1 again.

There is of course the difference that the oven temperature or water level is a continuous variable, whereas clearly the numbers generated by the algorithm are going to be discrete. We come back to this apparent difference, but for the moment accept it as similar.

The same algorithm could be written in a different but equivalent format:

1. Pick a number, call it y.
2. Replace y by the new number $y + 0.2 \cdot (1 - y)$ (this expression is indeed equal to $0.8y + 0.2$, so the algorithm is not changed.)
3. Repeat.

In this format the recipe reflects more what goes on in the capacitor. Each time the recipe is applied, the number y grows additively by a term $0.2 \cdot (1 - y)$. The present voltage level in the capacitor grows by an amount proportional to the difference between the supply level and the present voltage level. It stops accumulating exactly when y equals the supply level, which is 1.

Notice that for any initial y larger than 1, then $0.2 \cdot (1 - y)$ is a negative number, and so the number would decrease to 1. Although it is not a situation we described in the physical examples, it is easy to see that the situation can occur in these examples. If the water level was above the equilibrium to begin with (something that can be done by putting the plug in the drain and filling the sink before opening up the drain) than the outflow will be larger than the inflow and will drain the sink to such a level as to again have the outflow equal the inflow at which time the level will stabilize. Similarly if we had the oven filled with hot air above the temperature of the equilibrium for the oven, then more heat would transfer from the oven to the environment than the supply provided for, and so the thermal energy stored in the oven, and the temperature of the oven would decrease. This will go on till such time that the heat supply matched heat leakage, from which time onwards the oven temperature would again remain constant.

Our algorithm may be represented in yet another form, which reflects more directly how the numbers settle eventually at 1. To this end let us write the number that is obtained at the kth evaluation of our algorithm as $y(k)$. At the start, we have $y(0)$, then according to our recipe $y(1) = 0.8 \cdot y(0) + 0.2$, and in general at the $k + 1^{th}$ iteration $y(k+1) = 0.8 \cdot y(k) + 0.2$. The latter can be rewritten as $(y(k+1) - 1) = 0.8 \cdot (y(k) - 1)$. So at each iteration the difference between $y(k)$ and 1 is reduced by a factor of 0.8. It is now very clear that $y(k)$ will reach the value of 1, eventually[6].

---

[6] Strictly speaking, if we do not start with $y(0) = 1$ it will take forever to reach 1, but as computers cannot represent numbers to infinite degrees of accuracy, it will get there in a finite number of iterations or may never get there, and the sequence will hover around 1. You can try the experiment on any hand calculator, or excel spread sheet, start with any number and multiply by 0.8. After a few repetitions, it does not take long, the calculator (or spread sheet) will display 0 (i.e. the calculator does not implement multiplication completely accurately)!

2.5 How Similar Is the Behavior of These Example Systems?

It appears that completely different systems behave qualitatively and perhaps also quantitatively in the same way. Indeed many more examples are possible; like the speed a train reaches on the track, or the speed a fan reaches, or the height a helicopter flies at, and so on.

Qualitatively, the three physical systems expressed some form of conservation. In the oven, thermal energy is preserved; in the sink example, conservation of mass applies and in the capacitor, conservation of electrical charge is the key. In all these physical examples there was a single storage device (oven, sink, capacitor) and some form of compensation (heat leaked into the environment, water drained, or electric charge supply proportional to voltage difference). In the algorithm, we used a single number (a single memory location in the computer).

The similarities go beyond the qualitative. Indeed, with some care, we may construct ovens, or sinks, or batteries or algorithms that behave in identical ways from the signal point of view. What we mean is that we can use a specific scale for the time, and a scale for the temperature in the oven (or the voltage across the capacitor, or the numbers in the algorithm) in such a way that when presented in a graph, the water-level-in-the-sink signal cannot be distinguished from the oven-temperature signal (voltage signal, number signal). So from the signal point of view the oven is really a representation, or a different realization, or a model for what goes on in the sink and vice versa.

2.6 Discrete Time or Continuous Time?

There is no doubt the issue that the temperature in the oven appears to be a continuous variable, well-defined for at any time instant, as does the voltage across the capacitor or the water level in the sink and yet the numbers generated by our computer algorithm form a sequence of distinct numbers, not a continuum. In the latter case we speak of a discrete time signal, in the former of a continuous time signal.

True the temperature of the oven or the water in the sink can be measured at any time. In any observation though, there is always a finite resolution for how well we can measure the time (and the signal value) and so in any experiment, we will only collect finitely many different measurements. This process is called *sampling*. Also, how well we want to resolve time will depend on how fast the oven temperature or capacitor voltage or water level can change. The faster the variation, the better our time resolution has to be in order to get a good idea of the signal variation[7]. So from a data collection point of view, our computer numbers are more like the oven temperature, water level and capacitor voltage than the continuous curve we presented in Fig. 2.6. In fact, in any experiment the water level, or oven temperature would look like a set of data against the time for which Fig. 2.6 represents an interpolation, a continuous curve that links the observed points. Any form of *interpolation* is to some extent (intelligent) guess work.

When writing a mathematical description for the behavior of the oven, or water in the sink, we would use equations that we may glean from either thermodynamics

---

[7] There is a theorem about how fast samples have to be taken in order to capture the time variation of the signal, the so-called Nyquist-Shannon sampling theorem. We come back to this issue when we discuss signals in some more detail.

(the theory of thermal energy) or hydrodynamics (the theory of mechanics of water). Those equations are known as *differential* equations. In these equations time is a continuum and not discretely valued. Moreover it is assumed that the signals are functions whose derivatives are well-defined, that is they vary in a smooth way. There is however a simple way of converting these equations into equivalent difference equations, where time is represented by an increasing sequence of integers. In the difference equation representation, we look at the signal only at a number of typically equidistant instants of time. These difference equations are precisely like the computer algorithm we described previously. In many ways, whether we use discrete or continuous time has much more to do with convenience than essence, and we will use both, focusing on one or the other depending on which one is the more natural or most easy to use.

2.7 Analogous Systems

Our success in recognizing that these simple devices, seemingly underpinned by quite different physical phenomena, behave *analogously* leads us to the following understanding of what we want to call analogous systems.

> **Analogous Systems**
>
> Two systems are analogous if the collection of signals derived from these systems can be made identical by scaling.

In the representation of the signals we have of course a choice of how we measure time and signal values. This choice can only be exercised once, i.e. for each system and more particularly for the signals of each system we select a proper measurement device for the time and the signal value. Given the way we represent signals we now start comparing signals derived from different systems. If we cannot tell them apart (there is choice in what we actually call the same), then we accept the systems as analogous.

The concept of analogous systems is illustrated in Fig. 2.9.

One could object to the fact that system analogy is a concept relative to the observer and not an inherent system property. That is a fair point, nevertheless, we prefer it this way, as it captures the essence of how we interact with the world around us.

Moreover for most systems of any complexity it will not be possible to perform an exhaustive evaluation of all possible signals that systems could produce under all possible circumstances, which would be a requirement if we wanted a notion of absolute, observer independent analogy that could be a system property. In practice we always have to settle for less; that is a conclusion based on the experiments we have completed. Typically analogy will only be valid relative to our (collective) experience, but that is really all we have in science and engineering.

It is perhaps disturbing to realize that signals do not necessarily reveal the nature of the system. In our examples, whether it was a sink, or an oven, or even the computer algorithm one could not tell from the signal pictures, as these convey the same information in each case.

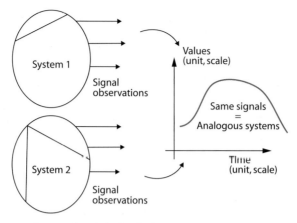

Fig. 2.9. Systems are analogous when their signals also are

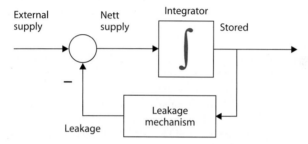

Fig. 2.10. Schematic view of a leaky integrator (applicable to any of the capacitor, sink, oven or computer algorithm example) as an example of negative feedback

This idea is very well captured by a block diagram representation for any of the systems we have described so far. Essentially we have a storage device (the sink, the oven, the capacitor, etc.) which we represent in the block diagram as an integrator. The integrator is characterized by one signal, its storage level. The storage is replenished, and some of its storage leaks away. The corresponding block diagram is represented in Fig. 2.10. The integrator represents conservation, or what is stored is the integral of the supply. In the block diagram the supply is input to the integrator and the storage is its output. If the supply is positive, storage increases, if it is negative, the amount stored decreases; if the supply is zero the storage remains constant. In the examples the total supply consisted of an external positive supply and a negative supply due to leakage. The latter is represented in the block diagram by the loop and the summation at the input of the integrator. The summation indicates that the supply to the integrator consists of an external supply and a leakage (negative sign at the summation junction). The loop in the diagram is known as *feedback*. In this case the feedback is denoted as *negative feedback*.

We will see in Chap. 5 that, in order to deal with a system, to represent, to analyze, to understand and/or explain its behavior, even to modify it, the concept of analogy is very useful. Analogies will be used freely to develop models of systems and, based on these models, the analogies will themselves become clearer.

2.8 Combining Systems and Splitting Distributed Systems

The behavior of a single oven, sink or capacitor is simple, both from a qualitative as well as quantitative perspective. The signals for a start are always monotone, either increasing or decreasing depending on whether the signal was above its equilibrium or below from the start. Oscillations in the temperature or water level in the sink will not be observed.

Somewhat more complicated behavior is possible when we start combining them together to make a new system.

For example we may combine a few sinks, such that the outflow of one is an inflow for another as in Fig. 2.11, not unlike a river or irrigation system. Or we may combine a few ovens together, so that there is thermal contact and heat may flow from one into the other. This is very much the situation in an industrial ceramic kiln, where tiles are manufactured (see Fig. 3.2). The tiles go through a number of different sections in the kiln, first preheating, then firing and finally cooling. Each of these sections of the kiln may be modeled as an oven (see Fig. 3.3) that interacts with its neighbors.

More signals are now required to describe the overall system. One temperature for each oven and one water level for each sink. But for most typical heat loss characteristics (where the heat loss increases with temperature difference) or drain characteristics (where outflow increases with available water level) the signals will eventually settle down. There will be a distribution of temperatures across the various ovens, but for each oven the heat supply will equal the heat loss and the water levels in the sinks will settle down such that for each sink separately the inflow will equal the outflow. The complexity of such systems has to do with the number of variables that we need to keep track off. Because of the above observed analogy, in principle, we could construct an interconnection of ovens, and a network of sinks that would behave in exactly the same way. It is however cheaper and more convenient and more flexible to use a computer algorithm instead.

Far more interesting behavior is possible if we extend the scope of the devices we interconnect (and admittedly few engineering systems consist only of 'leaky' sinks,

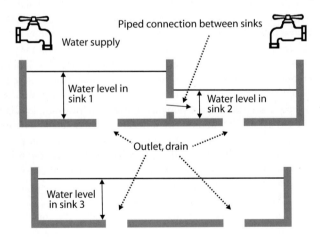

Fig. 2.11. A system of sinks

or ovens). If we extend the thermal world with cooling devices, valves to regulate mass transport, pipes, mixers, storage tanks and heat exchangers we get a far richer behavior. Such a system can be analogously implemented by using electrical devices (inductors, capacitors, resistors, diodes, voltage and current sources) or indeed mechanical devices (springs, inertia, dampers, force generators). The possibilities are endless. The mix becomes even richer if we also include chemical, nuclear and biological processes.

2.9 Oscillations

There are some properties which establish a clear difference between systems. One such property is stability (like the capacitor charge at equilibrium). We will deal with this concept in particular in Chap. 6. Another important property of systems behavior is the presence of oscillations.

To illustrate this interesting phenomenon, we will introduce what is the essence of oscillatory signals, or periodic signals, as well as oscillatory systems.

2.9.1 Energy Exchange

Oscillatory behavior is possible if we have systems where there is more than one type of energy, and where there is a simple mechanism to convert one form of energy into another. In mechanical systems we have potential (position related) and kinetic (speed related) energy. A falling object converts its potential energy (height) into kinetic energy (speed). In electrical systems we have electric (stored in capacitors) and magnetic (stored in inductors) energy.

Consider a mass suspended by a spring from a fixed point, subject to gravity. Let the mass be pulled down (by applying an external force) so that the spring extends. If the external force is now removed, we observe that the mass starts to oscillate. If there were no loss of energy (no friction in the spring) the mass will continue to oscillate between two extreme positions with equal distances from the point of suspension. During the motion the total energy remains constant but potential energy is continuously transformed into kinetic and back again. The motion itself is known as an *harmonic* motion. The idea is captured in Fig. 2.12.

2.9.2 Systems Perspective

From a systems perspective, we may represent the behavior of the mass-spring arrangement in Fig. 2.12 in a descriptive way by saying:

- The net force is null. Thus, the sum of the acceleration, friction and spring forces is zero.
- Acceleration force integrated gives velocity (Newton's law: force is proportional to acceleration, velocity is the integral of acceleration).
- The friction force can be assumed to be a function of the speed.
- The position is the integral of velocity.
- Force experienced by the mass is proportional to position (spring behavior, Hooke's law), and opposes further extension of the spring.

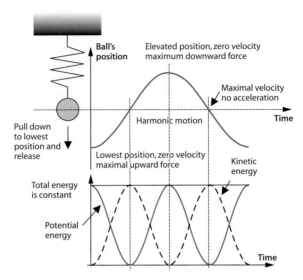

Fig. 2.12. A frictionless assembly of a mass and spring system. Total energy remains constant, but kinetic and potential energy are continuously interchanged

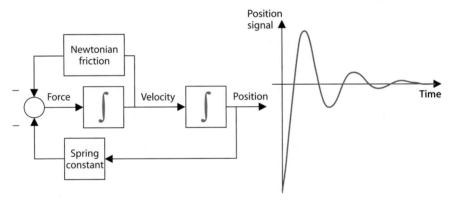

Fig. 2.13. A mass, spring system, with Newtonian friction

This is captured in the block diagram in Fig. 2.13. The interesting aspect of the block diagram is that it reveals negative feedback around a series interconnection of two integrators. The position signal can be seen in Fig. 2.12, it oscillates, assuming no friction at all. The velocity signal and the force signal also oscillate. The two integrators go hand in hand with the two forms of energy in the mass-spring system; kinetic energy (associated with velocity) and potential energy (associated with position).

A more realistic model of the mass and spring system may include some friction, for example Newtonian friction, which is a motion opposing force proportional with velocity. This is represented in the block diagram as a negative feedback loop around the integrator associated with velocity. With friction, the oscillations will eventually die out. A typical response is included next to the block diagram in Fig. 2.13.

2.10 Comments and Further Reading

The use of analogies is common in science and engineering education from the very first introductions to the topic, where the similarities are stressed through observation. The more formal analogy we have stressed here (systems are analogous when the externally observed signals are similar or indistinguishable, up to scaling of their index and value set) is really the back bone of systems theory and why systems theory is so widely applicable.

Moreover, often it is not essential to know what is in a system and as long as we can understand the interconnection and external signals we do have a valid explanation for the system behavior.

The approach to analogies, starting from signals, is inspired by behavior theory (Polderman and Willems 1998). The literature devoted primarily to signals, as in signal processing is vast. To cite just a few texts that put signals first, see e.g. McClellan et al. (2003), an introductory text, Mallat (2001), a more modern signal analysis approach based on wavelets, Ifeachor and Jervis (2002) devoted to digital, or sampled data signals. Signals are discussed in more detail in Chap. 4.

Analogy is also the foundation of bondgraph theory (see for example *http://www.bondgraph.info/*), which provides a common language to model all physical and in particular engineering systems (Breedveld and Dauphin-Tanguy 1992; Gawthrop and Smith 1996). There are simulation packages that provide convenient interfaces to quickly model engineering systems. There are also bondgraph-based toolboxes for modeling in symbolic manipulation packages.

Many classical texts on dynamics and modeling of engineering systems stress the analogies between electrical, mechanical and thermal processes (Cannon 2003). The main ingredient in these classical approaches is not the equivalence of the observed signals (as we do it here) but rather the observation that when writing mathematical equations to describe electrical, mechanical or thermal systems the same type of ingredients are used: conservation laws and constitutive equations, and when replacing certain variables from one domain by those of another the same equations reappear. This is what we would call classical analogy, as it approaches the modeling of systems from the point of view that the mathematical equations that describe the physical world around us are the starting point. Of course, these equations, an embodiment of the so-called physical laws are nothing else but a modern day representation of many past observations and measurements.

Let us conclude with a word of caution, cited from Cannon: *Analogies are detrimental, however, when they entice us into stopping our thinking about new physical phenomena ...* (Cannon 2003). In other words, when our models or analogies do not fit the data, we must adjust our preconceived ideas, familiarity cannot justify holding onto lies (and all models are lies to some extent). More about this when we discuss models in Chap. 5.

Chapter 3: Illustrating Feedback and Control

3.1 Introduction and Motivation
3.2 Manufacturing Ceramic Tiles
3.3 Gravity-Fed Irrigation Systems
3.4 Servo Design for Antennae for Radio Astronomy
3.5 Simple Automata
3.6 Homeostasis
3.7 Social Systems
3.8 Comments and Further Reading

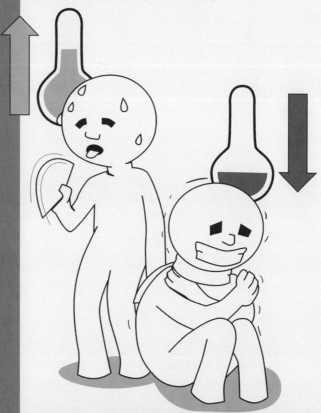

Chapter 3

Illustrating Feedback and Control

> *There is no free lunch.*
> Milton Friedman[1]

3.1 Introduction and Motivation

When there is a chill in the air, we feel uncomfortable and we are likely to put on some extra clothes and our body automatically reacts with shivers to increase our internal heat generation. If our environment is too hot, we sweat to extract heat from the body. This is feedback.

Typically, useful systems are complex and they derive their overall functionality from a set of different subprocesses that interact in a non-trivial way. This is the case even when the whole system serves a unique purpose. In nature this is the norm, and more and more engineered systems exhibit a similar structure and complexity.

The analysis and design of any system demands the integration of the physical understanding of the components' behavior together with control principles to achieve the overall purpose of the desired system. In design, the constraints matter most. The basic ideas underpinning the systems and control engineering approach to system design will be described in the following chapters. Here, a few examples are introduced to put the nature of feedback design into focus.

A system requires design at every level, from the scale of its smallest sub-system to the global scale involving the entire integration of all subsystems. A corresponding hierarchy of control levels, from local sub-system control to supervisory control strategies has to be designed to work seamlessly together in such a way as to achieve the system's goals and to sustain these over an extended period of time (see Chap. 10 for more details). Control cannot be an afterthought it must be an integral part of system design. Its fundamental limitations are just as essential as the perhaps more obvious physical constraints in the system. Equally control design cannot be meaningfully progressed without taking notice of the constraints imposed by the physics.

In this chapter we will illustrate the role of feedback and control to deliver purposeful behavior. Our examples are derived both from the engineered as well as the natural world.

We start with a description of a factory producing ceramic tiles. It serves as the archetypical manufacturing process, where the overall goal is reached through the execution of a sequence of discrete subprocesses. The ideas generalize to many manufacturing processes, at least in the structure of the global plant, being split into subprocesses with interconnections and interdependencies.

---

[1] Milton Friedman, US, Nobel Prize winning economist, 1912–2006, well-known for his contributions to consumption analysis, monetary history and theory, and stabilization policy.

Most natural and human made systems are distributed over time and space, hence the need for local measurements and local actions and a hierarchy to coordinate a global outcome. The control of large scale irrigation systems provides a good example of such a system.

The benefit of control in achieving accuracy and high repeatability are illustrated using the control of antennae for radio-astronomy. Accuracy combined with speed is the holy grail of mass manufacturing, perhaps best appreciated in the automotive industry.

Not all the control systems are so complicated. In everyday life we see how a car washing plant operates in a sequential way to (hopefully) wash, wax, rinse and dry our car. You can find many simple systems like this around us.

There are many regulatory processes in our body: the homeostasis of body temperature, chemical balances in our cells, blood and organs ... Normally we are oblivious to these processes, until something goes wrong.

Finally, feedback and control play an important role in social systems.

3.2 Manufacturing Ceramic Tiles

Process and manufacturing systems are easily split into subprocesses carrying out different logically ordered operations. Often the goal is to manufacture a particular product with its desired characteristics (quality), in the most economically viable manner. This objective translates into specific targets that have to be met at each stage of the process. The ideas presented here in the context of the manufacturing of ceramic tiles apply generically to manufacturing and chemical process systems.

In a ceramic tile factory the objective is to manufacture ceramic tiles that meet the expectations of the market in terms of quality, cost, durability and purpose. The production must of course be economically viable and the production process must meet all legal constraints.

In broad terms, the process lay-out is shown in Fig. 3.1.

The following subprocesses are identified:

- *Preprocessing* of raw materials. The different natural minerals (clay, lime, sand ...) are stored, pre-ground and moistened.
- *Milling*. The raw materials are ground and homogenized and mixed in the appropriate proportions, and stored in a slip tank.
- *Spray drying, and storage*. Milled and mixed material is spray dried and stored. This provides a buffer in the process.
- *Pressing*. The material goes from the slip tank to the press, where the "biscuit" is formed. Here the tile is shaped.
- *Drying*. Typically pressing is followed by drying, to reduce moisture content. Dried tiles can be stored before processing continues.
- *Glazing*. The upper surface of the tile is appropriately treated. Glazed biscuits are either stored (another internal buffer) or moved to the kiln.
- *Kiln firing*. In the kiln, the glazed biscuits are "fired". The tiles obtain their required mechanical and aesthetic properties.

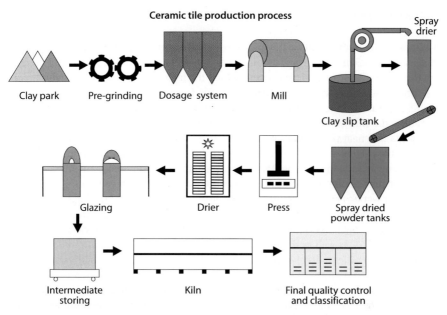

Fig. 3.1. Ceramic tile factory layout

- *Sorting and storing.* The final product, ready for delivery is stored. This is typically preceded by an inspection phase, in order to group tiles according to quality and consistency. Defective tiles are removed.

Production engineers maintain and tune the manufacturing process to get high production yields with consistent quality. The quality of the tiles depends on their dimensional accuracy, and general mechanical characteristics such as hardness and breakability, as well as their aesthetic appearance. The engineers tune the overall process to reduce costs and improve production flexibility at the same time. Cost reduction is mainly centered around saving energy, but also savings in raw materials play an important role, as well as improving the consistency. Other aspects of process tuning center around minimizing the environmental impact, reducing chemical and noise pollution.

Direct measurement of many of the important variables, like for example the mechanical properties of the tiles is difficult. Control objectives are as a consequence based on the more easily accessible process variables, which in turn affect the mechanical properties. Statistical quality control measures are used to try to minimize product variability.

Most of the final characteristics of the tiles depend on the firing process. The kiln is therefore the core of the whole production plant. The main control objective is to achieve a pre-defined temperature profile. In principle each tile should experience the same profile. This means that oven control is of critical importance. Also the conveyer belt speed determining the dwell time for the tiles in each oven section must be tightly controlled. Other variables that are maintained within specified ranges are airflow and air composition. The ventilation is of critical importance to ensure the surface finish of the tiles.

The Ceramic Kiln

A kiln is a long tunnel (can be 100 m or more) with a useable rectangular cross section of say 2 m by 0.5 m. The firing is a distributed process. The variables to be controlled are the air temperature and air pressure along the length of the tunnel. Both temperature and pressure must satisfy a specific profile to ensure tile quality. To simplify the modeling, but also remain in line with the limited freedom available in the control variables, the kiln is divided into particular sections. In some kiln sections combustion units will heat the air and the tiles. In other sections external air will be used to cool the tiles. Tiles are transported through the consecutive sections on a conveyer belt moving against the air flow. It is also important to distinguish between the upper and lower part of the oven cross section, above and below the tiles.

The kiln sections are functionally grouped in zones: pre-kiln, pre-heating, firing, forced rapid cooling, normal slow cooling and final cooling zone. The organization is presented in Fig. 3.2. The construction of the kiln, the number of sections, the distribution of the burners and fans, as well as the refractive material of the oven walls and the mechanical properties (rollers, drive speed) constrain the production possibilities, as well as the final product quality. The control goal in each zone is to maintain a temperature profile. The temperature profile is a function of the material that moves through the kiln. This is depicted in the upper part of Fig. 3.2. The control algorithm manipulates

- *the burners*, the flame temperature is controlled through fuel and air flow valves;
- *the fans* that adjust the airflow in the section, and determine the pressure and temperature in the section; and
- *the roller speed* to control the movement of the biscuits.

The control objective is to achieve the best possible tiles at the least cost within the physical limitations of the plant. Sensors[2] keep track of throughput, air temperature,

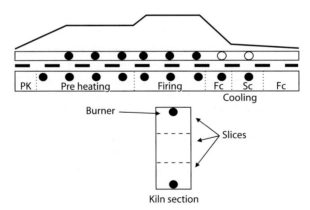

Fig. 3.2. Schematic view of a continuous kiln used to bake ceramic tiles

---

[2] See Sect. 9.2 about sensors and data acquisition systems.

fuel flow and airflow. The physical construction of the oven enables the control, and in order to be really effective, the construction and control design are to be coordinated.

In order to optimize the quality of the tiles at the kiln output, two kinds of control are used. A local control loop keeps the temperature profile along the kiln based on temperature sensors in the walls of the kiln. A supervisory control, rule-based, modifies the desired temperature and pressure profiles to counteract quality defects that are measured at the kiln output (Bondía et al. 1997).

The common sensing devices are thermocouples for temperature, and pressure and speed sensors. The actuators are the motors that manipulate the rollers and fan speed, and the valves that modify the flow of fuel and air.

On exiting the kiln some tile properties can be measured like dimensions, shape (flatness) and color. Mechanical strength can only be measured off-line. Typically this involves a great time delay and a destructive test, therefore this is not suitable for on-line control of the production process. A large number of variables are measured for ongoing monitoring purposes.

Kiln Control

The control goals for the kiln are to guarantee the quality of the tiles, to speed up production and minimize costs (energy, waste). To achieve these (competing) goals, the following controls are available:

- roller speed,
- temperature profile,
- pressure and ventilation profiles,
- transition between biscuit batches and
- batch packing density.

Most of these can be set for normal operational modes, and local control ensures then that the *set points* are maintained within acceptable tolerance margins. It is more difficult to create the entire collection of appropriate operational conditions. For example, some defects in the tiles may actually be traced back to the pre-processing (mill, press, dryer) and some of these can be corrected, perhaps partially, by appropriate action in the kiln. In order to minimize waste it pays to continue the production and indeed modify the operation of the kiln to compensate for the defects in as much as possible. When the kiln is operated manually, experienced operators know intuitively how to change the temperature profile or ventilation profile to drive the kiln to a new steady state to take these appropriate corrective actions. In automatic mode, however, these type of corrections are particularly difficult to carry out, and normally one has to resort to rule-based control actions, whereby the actions of *experienced* operators are *mimicked*.

The following hierarchical control structure is usually implemented

- Over the life span of the kiln the supervisory level of control continuously monitors the main variables, logs and acts on alarms, stores process data for off-line analysis and further improvement of the exploitation of the facility and an appropriate man-machine interface. Economic analysis of the product in the market place is linked

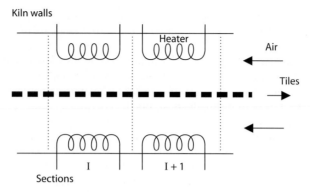

Fig. 3.3. Schematic view of the kiln: tiles move to the right and air flows to the left

with design and production, to provide the economic drivers for further investment in the production facility.
- On the production planning horizon, batch planning involves determining the kiln profiles, the packing density and scheduling of the kiln to ensure minimal transients between different production profiles. The planning provides set points for the burners (fuel flow, air flow), roller speed, and ventilation.
- Over the duration of a single production batch, quality control-based on measurements at the exit of the kiln is used to fine-tune the temperature and ventilation profiles. These changes are decided using heuristics or table look up methods.
- Over the duration of a particular batch the desired temperature profile, ventilation profile and speed profile are maintained using local controllers, which act in each of the separate kiln sections (see Fig. 3.3). The control actions are coordinated in such a manner as to maintain the section at the right set point whilst minimizing interactions with neighboring sections.
- For each actuator, a local controller provides the desired command based on their sensory input and the reference signal. The reference signals are provided from the quality control supervisory algorithm and the coordination control layer.

Producing Good Tiles

To produce quality tiles you need

- quality raw materials,
- excellent tile design,
- excellent processing units (mills, presses, kilns, ...)
- skilled operators,

but you also need

- precise control units and control algorithms,
- integrated factory information,
- quality control,
- automatic process management.

3.3 Gravity-Fed Irrigation Systems

Most utilities, like electrical power generation and distribution, water harvesting, bulk transmission and distribution and gas distribution are large scale engineering systems requiring careful control in order to match supply with demand, when and where it is needed. Similar control issues arise in the control of transport networks and even the internet or telecommunication networks more generally. The challenging issues here are the geographic scale and the variety of time scales on which dynamics are important. In such applications control is invariably organized to be distributed over space as well as hierarchical to cope with the many different time scales. The example we discuss relates to the control of irrigation networks, where the implementation of distributed control schemes is a recent phenomenon. The discussion will emphasize the spatially distributed nature of the control on the fastest time scale. The issues of supervisory control responsible for scheduling, long term exploitation and (preventive) maintenance as well as alarm handling are equally challenging and important but not that different from any process control application, like the ceramic tile factory.

In Australia irrigation accounts for 70% of all fresh water usage (UNESCO 2006; The Australian Academy of Technology Science and Engineering 2002). The main civil infrastructure for irrigation consists of reservoirs for harvesting and storing water and open canals for water distribution. The distribution of water is controlled through regulating structures that can restrict the flow in the canal between virtually no flow and a maximum flow which depends on the geometry of the canal (slope, and cross section) and the available water head (potential energy) at the reservoir feeding the canal. Here we look at how large scale distribution of water powered purely by gravity can be automated with great advantage above typical manual canal operations.

When water is in ample supply, and hence a cheap commodity, there is no economic pressure to be efficient in water distribution, and one way to operate the canal infrastructure is to ensure maximum flow, which guarantees the best water availability to the users. Water distribution is then a mere scheduling problem, and there is no need for closed loop control. Scheduling, which is a form of supervisory control, is unavoidable as typically the combined flow capacity of all water outlets onto farms is about ten times the flow capacity at the top-end of the canal system. Scheduling ensures that demand is averaged out to stay below the flow capacity of the canal system and such that the water is distributed to all users in an equitable manner. When there is an over-supply of water, this water simply returns back to the natural river systems or ground water storage, and is no longer available for irrigation (in that same season). Although present manual operations are certainly not operating on this maximum flow principle, which would be maximally wasteful, present canal exploitation reportedly achieves at best between 55% to 70% water efficiency and in many places around the world no better than 50%. Moreover a four day in advance water ordering policy is enforced in most irrigation districts, which does not favor efficient irrigation on farms, as farmers have to minimize the effect of uncertain irrigation timing. About 10% to 15% of the water is wasted through seepage and evaporation (no amount of control can change anything about that of course) and about 20% to 30% is lost through outfalls or is unaccounted for (The Australian Academy of Technology Science and Engineering 2002).

Policy makers in Australia have recognized that it is important to reconsider present water practices in view of long term environmental and sustainability issues. They are providing clear economic incentives to be water efficient, both in canal exploitation and on-farm water usage. Climate change, population and industrial growth pressures compound the sustainability problem.

Irrigation efficiency, whilst maintaining the other objectives of water level regulation (which is the potential energy available for irrigating farm land) and meeting water demand is an ideal objective for closed loop control. The conflicting requirements between meeting demand and achieving water efficiency make for an interesting challenge.

In order to realize closed loop control the existing civil infrastructure has to be upgraded with an information technology infrastructure of sensors and actuators linked through a Supervisory Control And Data Acquisition (SCADA) communication network. The approach to efficient water distribution is in three stages: building the information infrastructure, extracting the data to build models for control and finally closing the loop. This infrastructure enables automated decisions to set the regulating structures such as to deliver the water, and only this water that has been requested by the farmers. A project of this kind has been ongoing in Australia from about 1998, and it has achieved pleasing outcomes with canal operations running at around 85% efficiency, achieving a high level of on-demand water delivery (more than 90% of water orders are not rescheduled) and maintaining excellent water level regulation. In addition, because the system is now more responsive, farmers have adopted irrigation schedules that are much better suited to the crop needs with additional and significant water efficiency improvements and better economic returns on-farm.

The information infrastructure underpins the entire control approach, and can support decision making on all time scales ranging from hours to years:

- On the longest time scale the main issue is sustainability: how to best use the limited renewable water resource. This involves the development of appropriate infrastructure, policy and pricing mechanisms.
- On a yearly and seasonal basis the allocation of water volumes and crop plantings/treatments are decided according to the specific economic and also environmental requirements. The existing infrastructure is maintained and upgraded.
- On a weekly basis irrigation is planned to meet needs (on this time scale the local weather forecast plays an important role), and bulk distribution is scheduled at this point.
- On a daily time scale water is scheduled into the down-stream end of the canal system.
- On an hourly time scale individual canal sections under control react.
- On a minute time scale, water levels and flow regime are regulated and hardware and variables of interest are monitored to enable preventive maintenance and to ensure a graceful degradation in performance when sensors, actuators, or the radio network develop failures.

Figure 3.4 provides a picture of the information infrastructure hardware put in place in Victoria, Australia in juxtaposition with the old mainly manually operated technology that it is replacing. At regulation points, water levels are measured and water flow

Fig. 3.4. Radio networked actuator/sensors for irrigation, developed in Australia by Rubicon Systems Australia, in collaboration with the University of Melbourne (**c**) replacing the old manually operated infrastructure (**a–b**)

is inferred. Through communication with their peers over the entire network, a real time water balance can be deduced. Besides these main variables, a host of other variables is measured for maintenance and operational purposes (like battery levels, solar radiation, sensor calibration, etc). A schematic of the radio network, which allows for peer-to-peer and broadcasting, is represented in Fig. 3.6. Any regulator, with its associated actuators, and sensors has an internet address and can communicate via the radio network with any other regulator in the network. Most communications are based on an exception protocol, a communication is initiated only when something interesting happens. Regular polling is also performed to interrogate (parts of) the entire system to establish health checks. Broadcasts are performed to facility software upgrades, coordination and general management of the networked resources.

Data derived from the sensors and actuators is used with appropriate model parameter estimation and system identification techniques (Weyer 2001; Eurén and Weyer 2007) to develop simple models that relate control actions (at up and down stream regulators) to water level (and flows) on the basis of a single pool, that is a stretch of canal between two regulating structures. The emphasis is on simplicity, as irrigation systems are large scale and hence the models must be able to grow with the system. To get an appreciation of size consider the Goulburn Murray Water district, with more than 6 000 km of irrigation canal, more than 20 000 customers and 5 000 regulating structures spread over an area of 68 000 km^2. (There are much larger irrigation systems.)

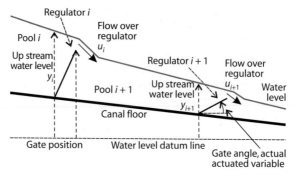

Fig. 3.5. In-line canal system, a series of regulators and pools (sections of canal between regulators)

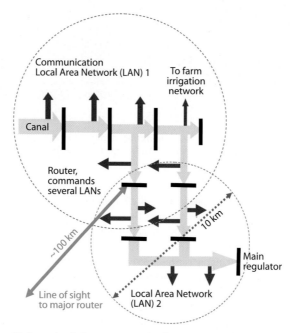

Fig. 3.6. Diagram of information infrastructure and radio network for an irrigation canal

To focus the ideas, we consider the smallest time scales and consider a simple canal, with a number of consecutive pools like in Fig. 3.5. A very simple model could be:

```
Change in pool volume = (in flow - out flow) X time_interval
```

It only captures water storage. This is a little too simple, and see for example Weyer (2001) and Eurén and Weyer (2007) for more comprehensive models and discussion. Nevertheless, it is enough to get some appreciation for the problem. The canal model is then the collection of all the pool models combined with the models for the regulators, all of which are derived from the measurements, and maintained in a central data base.

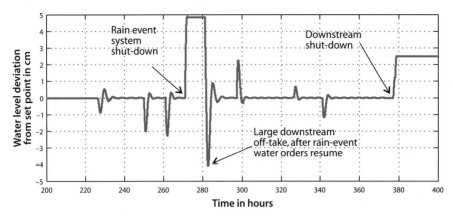

Fig. 3.7. Control response to a significant rain event

The control objective is to ensure that the water levels are regulated at the desired levels whilst the control inputs are constrained to be within their physical capacity limits (between open and close). All of this despite significant disturbances in the form of water off-takes, evaporation and seepage. Moreover it is important that the demand placed on the radio communication is limited and that the irrigation network copes with inevitable communication errors, as well as hardware failures in sensors and/or actuators.

The control objective (no losses, water level regulation, meeting demand) is achieved using a two staged approach. When a water order is placed (e.g. through the internet), the central node verifies against a global model of the system if this water order can be delivered as requested within the system capacity. It also checks prices, and water allocation rights, but that is irrelevant for the control process. Here the use of demand forecast including a weather forecast can be used to advantage to manage the irrigation district. If the water order can be delivered then the central node informs the canal regulators and the on-farm off-take gate of the requested water order. The flow is then implemented and the local controllers ensure that the water level is maintained, despite the imposed flow changes.

A typical response of the controlled system, in a week of operations on a particular pool, is presented in Fig. 3.7. The figure shows the quality of the set point regulation (deviation measured in cm). Under manual operation, the deviation from the desired level was considered adequate if the water level was within 25 cm of the desired level. The rain event, a serious disturbance, causes a shut down of the system as all irrigators on the channel stop irrigating. After the short rain event major water orders are resumed. At the end the effect of saturation can be seen, as the gates close in view of no down-stream demand.

Similarly in Fig. 3.8, a single day of operations is displayed, and both flow and level responses are presented. Notice the enormous flow variation over the one day period. This demonstrates the flexibility of the automated system.

Field trials over a number of seasons have demonstrated that the automated system, called Total Channel Control^{TM}, is very effective (Cantoni et al. 2007; Mareels

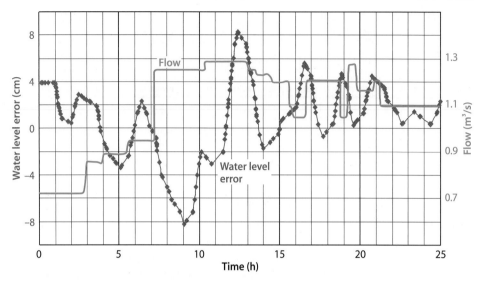

Fig. 3.8. Control response to large flow variations

et al. 2005). The automated system brings a number of side benefits that are not to be underestimated:

- more precise accounting of water flows and water volumes (order of magnitude improvement in flow metering accuracy),
- leak detection (water balances can be maintained per pool, pools requiring maintenance are easily identified),
- redundancy in data and sensor hardware can be used to reconfigure control action in case of hardware failures, ensuring a very graceful degradation of system performance under hardware failures.
- water ordering flexibility, which leads to better on-farm water efficiency.

Further developments include

- integration of weather forecast into the feedforward in order to improve the reaction to rain events, and manage system wide storage and flow conditions,
- flood mitigation (this is essentially reservoir control, using the canal system to disperse excess water),
- integration with river control, optimizing the use of the water resource.

The final goal being the real-time closed-loop management of water resources on the scale of an entire water catchment area (Mareels et al. 2005). This appears entirely feasible, and would go a long way towards the management challenge raised in the UNESCO World Water Report 2003, repeated in the 2006 and 2009 editions of the report (UNESCO 2006; 2009), all of which talk about a water management crisis in the world.

Best Use of Water

Water is a limited renewable resource, in need of (better) management that requires

- accurate information about resource availability,
- clear management policies, appropriately prioritized,
- substantial civil infrastructure (reservoirs, canals, pipes, valves),

but also

- real time water (level and flow) measurements,
- demand measurement and/or forecasts, and short term weather forecasts,
- wireless communication network,
- process models, these can be derived from the measurements,
- control design enabling in conjunction with policy, optimal exploitation of available resources within physical constraints.

3.4 Servo Design for Antennae for Radio Astronomy

Antennae are everywhere: they capture particular radio waves, that have been encoded to carry information. Often radio waves are directional and antennae orientation relative to radio wave is critical for a good reception, and subsequent information extraction.

Large antennae as for example used in radio astronomy and satellite tracking applications achieve the alignment between radio waves and antennae structure through positioning the antennae surface accurately in both elevation and azimuth. The precision with which these angles have to be achieved in modern radio astronomy is extremely high (errors are measured in arcsec, of which there are 3 600 in a degree or 1 296 000 in a full circle; an arcsec is approximately 4.85 microradian).

In the case of radio astronomy the reference trajectories for the elevation $\theta^r(t)$ and azimuth $\phi^r(t)$ angles that describe the path of the object(s) to be tracked (or points of the sky to be explored) are typically provided by the operators of the facility to the control unit as an ordered list of timed reference angles (time series) in a table:

$$(t_k, \theta_k^r, \phi_k^r) \quad \text{for } k = 0, 1, 2, \ldots$$

where t_k are the specified times, θ_k^r the required elevation angle and ϕ_k^r the required azimuth angle at that point in time. This table is to be interpolated over time to arrive at the actual reference angle as a function of time, (the angle has to be a continuous function of time). For example using linear interpolation (this is too simple in general, but used here to illustrate the idea) and with the above list the reference for the elevation angle becomes the following piece wise linear function of time, as displayed in Fig. 3.9.

The servo control problem is to ensure that the actual pointing angle of the antennae dishes follow the prescribed path with great accuracy despite disturbances such as a variable wind load (Evans et al. 2001). This is quite challenging because typically the drive system as well as the antenna structure exhibit multiple, poorly damped, resonant modes. Moreover, there is a wide variety of reference trajectories.[3]

---

[3] To learn about system frequency properties and resonances go to Sect. 5.7.2.

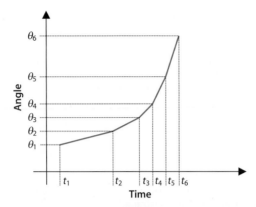

Fig. 3.9. Piece-wise linear interpolation of data points

Fig. 3.10. Some of the antennae in the Australia Telescope Compact Array, Culgoora, New South Wales, Australia (© Commonwealth Scientific and Industrial Research Organisation, CSIRO)

Figure 3.10 displays some of the antennae of the Australia Telescope Compact Array (ATCA); for more information visit *http://wwwnar.atnf.csiro.au/*. The antennae in this array were used for some of the experimental results described below. They have a diameter of about 22 m and weigh about 60 tons.

The control of large fully steerable antennae has been the subject of practical and theoretical studies for over five decades. The actuation and sensing subsystems pose significant challenges and determine to a large extent the limits of accuracy that feedback design may achieve.

The key mechanical design questions pertain to the maximization of the structural resonances (Wilson 1969) whilst trying to keep the cost of the structure low. The tradeoff is to achieve a high mechanical stiffness with a low overall weight, so the antenna can move faster and work harder for the radio astronomer. The lowest structural resonance frequency must be sufficiently high so that normal wind conditions cannot induce vibrations that render the antennae useless. The antenna is very much a large sail. Also, the lowest structural resonance frequency should be larger than the fastest angular velocity of the reference trajectories that are to be tracked with precision, otherwise the references would excite these resonances[4].

Most of the wind energy is found below 0.5 Hz. See Fig. 3.11.

The overall control design is greatly assisted by using high gear ratios (gear ratios of the order of 40 000 : 1 are not uncommon), such that the entire inertia is dominated by the drive motor's rotor rather than the wind loaded dish. This means that accurate positioning of the rotor of the drive motor implies accurate positioning of the antenna dish provided of course that the structural resonances are not excited. The drive train resonances can be actively dampened using feedback control.

The backlash[5] (see Fig. 3.12) in large gear trains is unavoidably large. Backlash is due to a necessary gap between the gear teeth of the driven and driving gear. When the motor force reverses, there is a period of time that the teeth do not mesh; so the motor moves but the driven gear does not move until such time the gear teeth mesh again. Nevertheless, the negative effect of the backlash on the tracking accuracy can essentially be eliminated

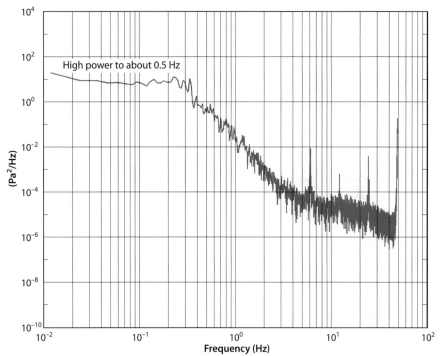

Fig. 3.11. A typical power spectrum for wind load on a large antenna structure

---

[4] See Sect. 4.3 for a definition of power spectrum.
[5] Backlash is a non-linear effect, which appears whenever there is a gap between some mechanically coupled components.

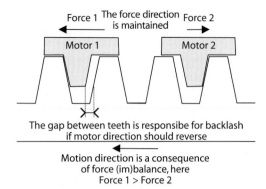

Fig. 3.12. Backlash and backlash elimination by dual drives in large gear trains

through the use of dual drive motors that work in a torque-share-bias mode. At each motor the force direction is maintained, so that one motor drives, whilst the other one is actively braking, as illustrated. Because the motor forces never reverse direction, the gear teeth mesh at all times, and the effect of backlash is all but eliminated[6]. All the antennae used in the experimental results below were built along these lines. The main limiting factor remaining in the drive train is stiction, which is a consequence of the large mass supported by the gear train. Stiction fundamentally limits how slow the antennae can move. The motor drive must provide a minimum imbalance before the motor forces induce motion. Once motion starts, the friction force drops dramatically creating a speed up effect being balanced by a reduction in force and consequently stiction takes over, the process repeating itself creates a limit cycle[7]. In the Australia Antenna Telescope stiction induced limit cycles consume 50% of the position error budget.

To achieve accurate tracking, precision sensors for the actual pointing angles of the antennae are required. Measuring the pointing angle of a large antenna is non-trivial. Recent antenna designs employ high accuracy position sensors, like 22 bit angle encoders that resolve angle position to within 0.3 seconds of arc.

The resonance frequencies of the antenna structure depend on the actual dish position, a low elevation angle will result in a lower resonance frequency. This is illustrated in Fig. 3.13. It shows how the resonance frequencies vary with the elevation angle in one of the Australia Telescope antennae. More generally, the resonance frequencies are a function of the geometry of the antennae, and every time the structure is updated or changed these resonances change as well. This implies that the control strategy must be able to cope with these variations. In a general setting, that means different process behavior in different modes of operation, requiring different control actions.

Following the experience with the Australia Telescope (Evans et al. 2001) several alternative servo designs were completed and tested on different large antennae. Under the assumption that the gear train has been properly designed to minimize stiction

---

[6] Nature makes extensive use of such arrangements. Two components of opposing effect are present at all times and their balance determines which effect dominates.

[7] A limit cycle implies maintained oscillations around an equilibrium point. This is typical in forced oscillators and it only appears if the system behavior is non-linear.

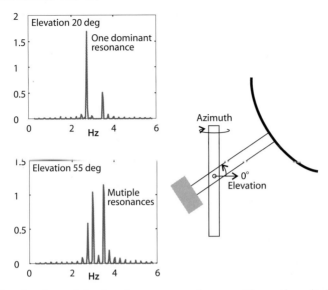

Fig. 3.13. A low elevation angle implies a lower resonance frequency. Observe the multiplicity of resonances as an indication of the complexity of these antennae structures

and that a back lash compensation drive scheme is employed, a cost effective servo design consists of four hierarchically organized feedback loops (see Chap. 8):

1. a very fast, high gain, current feedback loop, such that from a tracking perspective the motor drive is a pure torque source. This loop operates up to about 100 Hz.
2. surrounded by a medium bandwidth[8] velocity feedback loop, which dampens the main gear resonances. This loop is effective to about 10 Hz.
3. enclosed by a low bandwidth position tracking loop, which ensures set point regulation whilst avoiding resonance vibrations. This loop is active to just under 2 Hz, just below the slowest resonance frequency. (Notice that set point regulation is not enough, as the reference trajectory is not a constant, but a function of time!)
4. finally, the slowest outer loop minimizes the remaining tracking error. This loop effectively compensates for the variation in the resonance modes by adjusting a feed forward gain from the reference trajectory. Moreover it ensures that arbitrary reference trajectories, not just constant functions of time, can be followed with acceptable tracking error. It is called an *adaptive loop*. This loop operates to about 1 Hz, about twice the typical bandwidth of the wind disturbance, and well above the required rate for the reference signals (stars do not move that fast across the sky!).

Such a control structure, is called a *cascade*. The idea is illustrated in Fig. 3.14. The above discussion focuses on the local control objective for the radio antennae. There

---

[8] Formally, the bandwidth refers to the range of frequencies a signal is composed of. In colloquial terms, the larger the bandwidth the faster the signal changes. If referred to a process, for instance a loudspeaker, its bandwidth denotes the range of sound frequencies it reproduces with good fidelity.

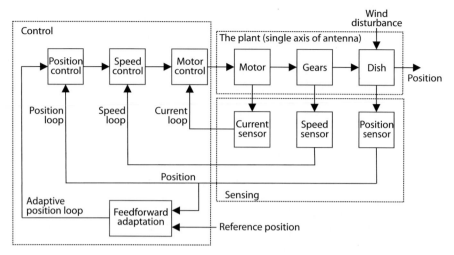

Fig. 3.14. The servo mechanism's control structure: four cascaded loops. The external input signals are the reference position and the wind disturbance

are however other aspects that need to be catered for in any operational control system for such antennae:

- Scheduling of the experiments, this provides the reference trajectories for all the objects to be tracked, as well as the start and stop times for each of these.
- Supervisory control, to start and stop the operation of the antennae, as well as interrupting normal operations and stowing the antenna when the wind load is outside the antenna's operational envelope or when faults are detected in the drive or sensor units.
- Tracking mode, as described above, where the antenna tracks a given reference over a set period of time.

> **Observing the Sky**
>
> Watching space requires precision instrumentation and infrastructure:
>
> - large antennae spread over a large geographic area (the radio telescope),
> - precision communication between antennae to coordinate these,
> - skilled operators,
> - appropriate and precisely surveyed locations (no radio interference),
> - excellent computing facilities,
>
> and the antennae themselves must be complemented with
>
> - precision mechanics (drive train),
> - precision instrumentation, inter alia measuring position,
> - dynamic models of antenna behavior and wind load conditions,
> - control design to track a large range of signals (stars) under a wide variety of wind conditions (robust control).

3.5 Simple Automata

There are many common processes where the overall behavior can be adequately captured using only binary valued signals (on/off, stop/go, open/close). In such circumstances it is often possible to implement the control also only using binary valued signals. An important class of such systems is called the *automata*[9]. Examples range from a simple vending machines to the complexity of a digital computer.

Just to illustrate the idea, let us consider a car washing system, as represented in Fig. 3.15.

The system operation starts when a coin or a token is introduced in the machine slot. The sensor M turns on the green light LV. The system is ready to operate as soon as the user clicks on P. This action will turn on motor $C1$. The belt will move the car until it reaches the positions sensed by $S0$. This event will start the motor $C2$, activating the displacement of the attached belt and turn off the green light. The car enters the washing area. Its presence is detected by the sensor $S1$. The car is then soaked with soapy water dispensed through the pump AJ. After a predefined time $t1$ the front washing rollers start to work (motor $MR1$).

The sensor $S2$ will detect the approach of the car and will start the following section, activating the belt $C3$. Once the car is in the new section, the green light is turned on allowing the entrance of a new car into the car wash. Sensors W_i are placed at the exit.

A similar process is repeated in the rinsing and drying sections.

A number of additional binary signals verify the overall operation to prevent accidents from occurring. In particular, there is always an emergency stop button that the user may activate if something untoward is happening.

Factory automation is often much like this example, only more complex. Home automation like in a dish washer or washing machine is also of this form.

> **Process Automation**
>
> Factory and process automation are common.
> Even complex automation can often be restricted to binary signals, for analysis, design and implementation.

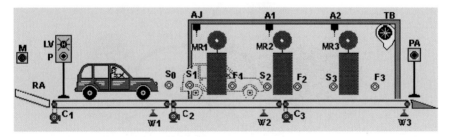

Fig. 3.15. The automated operation of a car wash

---

[9] Automaton: a device or machine designed to follow automatically a predetermined sequence of operations. The modern analysis of automata can be traced back to John von Neumann and Alan Turing.

3.6 Homeostasis

Homeostasis is a term used in the context of biology, with reference to maintaining the same desirable condition. Indeed, life as we know it depends on the existence of a well-regulated chemical and physical environment inside the cells. This chemical and physical balance is maintained through active feedback mechanisms. In most examples of homeostasis we can distinguish all the functions we find in an engineered control application: sensing, communication, signal processing and actuation. In the body the control sub-system is an integral part of our body to the extent that usually we are not consciously aware of this activity, apart from those times when we become sick. There is a huge variety of interacting feedback loops in the human body. Feedback is at the core of such vital things as our body temperature, the amount of oxygen in our blood, the amount of glucose (the fuel we use in our cells) and so on. There is at least one feedback loop for almost any particular chemical that exists in any of our cells.

It is not surprising to find these feedback loops well organized according in a hierarchy of feedback loops, just as in engineered systems. There are local, fast acting, control loops that maintain certain variables within acceptable margins. There are supervisory control actions that monitor vital symptoms and kick in secondary feedback mechanisms in case of major trauma (like bleeding, dehydration, overheating) or abnormal conditions due to failures in sensing or actuation, or because of bacterial or viral invasions. Finally, there is the conscious decision level, where we decide things like what to eat or drink, or when to exercise or go to sleep all of which have a major impact on our body functions and require specific intervention by all the lower level control loops in order to deliver homeostasis.

By way of example we consider somewhat superficially the condition of diabetes mellitus, when glucose regulation fails. This section is adopted from Santoso and Mareels (2002) and Santoso (2003). The word diabetes is derived from Greek and means as much

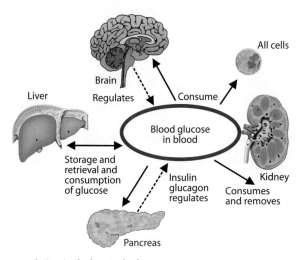

Fig. 3.16. Glucose regulation in the human body

as *passing fluid* and mellitus is Latin for *honey-sweet*. The word diabetes was coined in AD 130 by Aretaios of Cappadocia in his book about the disease. Excessive and sweet urine is indeed a characteristic of having too much glucose in the blood and no appropriate insulin mechanism to control it, which triggers a secondary feedback through the kidneys to remove the excess glucose from the bloodstream. The first reference to diabetes is found in Egyptian literature from 1550 BC. Other early references are found in Chinese, Japanese and Indian writings. The disease did not receive an effective treatment till the discovery of the insulin treatment by Banting and Collip in 1922.

Glucose is the most important fuel in the human body. All our organs, not to mention our brain require the right glucose supply to function properly. Glucose is transported to the cells via the bloodstream. Maintaining the right amount of glucose in the blood stream

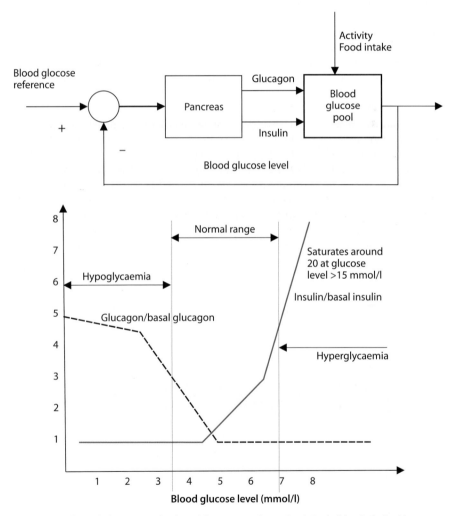

Fig. 3.17. Insulin and glucagon production with respect to glucose levels in the blood of a healthy person

is achieved through an intricate feedback. Too little glucose (hypoglycaemia[10]) and our body stops functioning normally (loose consciousness for example), and too much glucose (hyperglycaemia) leads to the awful late-onset complications diabetes is feared for (especially when our cells are exposed to it over prolonged periods of time).

The supply of glucose is derived from our food, which in our digestive system is re-manufactured into glucose $C_6H_{12}O_6$. The release and storage of glucose is regulated through hormones (see Fig. 3.16). The brain acts as the main controller and the liver acts as the main glucose storage. Glucose enters the cells through their surface requiring a glucose concentration gradient being higher in blood than inside the cell. *Insulin*, a hormone produced in the pancreas, is responsible for re-moving glucose from the bloodstream. It acts on the cell's membrane allowing the flow of glucose as well as storing any excess glucose in the liver (and some gets stored inside our muscle tissues as well) in the form of glycogen (a polymer derived from glucose) for later use. Extra glucose is flushed out in urine through the action of the kidneys.

The pancreas also produces the hormone *glucagon*, regulating the amount of glycogen. If more energy is required, glycogen is converted back into glucose and released into the bloodstream. The pancreas maintains a base level of insulin and glucagon in the bloodstream, regardless of blood glucose levels (remember the dual drive in the antenna Sect. 3.4).

The pancreas increases the insulin in the bloodstream when the glucose levels rise above 5 mmol/l and similarly increases the glucagons level when blood glucose level drop below 4 mmol/l (see Fig. 3.17). Physical or mental activity also influence the glucose level, mainly determined by the intake of food. In healthy individuals blood glucose is maintained through this mechanism within the range of 3.5 to 7 mmol/l. Less than 3.5 mmol/l is considered hypoglycaemic and more than 10 mmol/l is considered hyperglycaemic. In a healthy individual, the level of glucose varies with physical activity and food intake. A typical 24-hour record of blood glucose is displayed in Fig. 3.18.

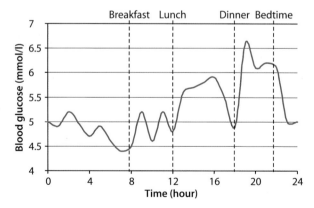

Fig. 3.18. A blood glucose signal of a healthy person

---

[10]This is derived from the Greek, glykys = sweet and haima = blood and hypo = below.

When suffering from Type I Diabetes Mellitus, the pancreas does not produce enough insulin and the presently prevailing treatment consists in providing insulin through subcutaneous injections in response to measured blood glucose levels, expected food intake (more food requires more insulin) and levels of activity (more activity requires less insulin).

Compared to a normally functioning pancreas, a regime of regular subcutaneous injections (see Fig. 3.19) is at a serious disadvantage to regulate glucose. The pancreas monitors the blood glucose level continuously, in closed loop, and applies near instantaneous feedback. In the injection regime, the glucose level is in open loop behavior for extended periods of time (between injections). A typical rule-based control algorithm, as implemented by the patients, on how to decide how much insulin to inject is graphically represented in Fig. 3.20.

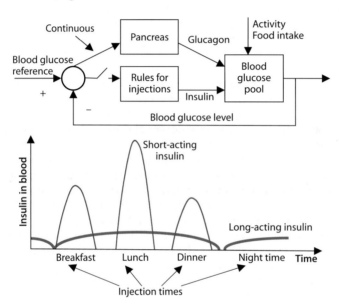

Fig. 3.19. Insulin injection regime using slow and fast acting insulin preparations

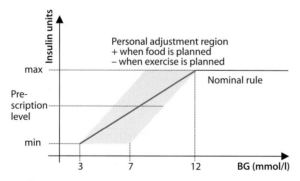

Fig. 3.20. Insulin injection rule, showing how an individual may vary the recommended dosage of slow and/or fast acting insulin

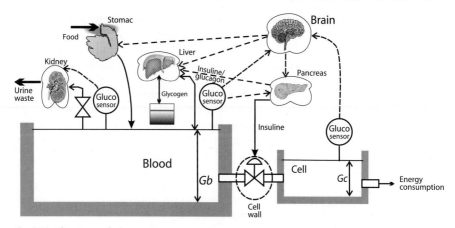

Fig. 3.21. Glucose regulation

Figure 3.21 illustrates the glucose regulation. The main goal is to regulate the level of glucose inside the cells, Gc, and this level depends on the cell's workload. The brain (and the sympatic nervous system acting as the central controller) receives demands from the cells as well as information about the glucose level in the bloodstream, Gb. This glucose level is also sensed by the pancreas to determine the production of insulin or glucagon. The kidney has a rough sensor of the glucose level (by osmosis) to extract excess glucose. Under normal operation, any time food is processed, the level of glucose increases, the pancreas generates insulin and glucose starts to be stored as glycogen in the liver as well as muscles. This glycogen will be used to regenerate glucose in periods without food intake, or whenever extra energy is required.

The cell membrane can be considered as a one-way valve (see actuators in Chap. 9). The insulin acts as to open the valve allowing glucose into the cells. Diabetes is here illustrated as an inability to open the valve.

The actual system is much more complex. A more careful consideration of glucose regulation will reveal a plethora of different feedback loops organized in a hierarchical structure.

> **Homeostasis and Feedback**
>
> Our body relies on feedback. From motor activities, such as standing, walking, grasping, reading a book, to any other human conscious and unconscious activity (e.g. metabolic) feedback plays a fundamental role.

3.7 Social Systems

Feedback plays an important role in our society, recall for example the teacher-student example from the Introduction, see also Fig. 1.1. Feedback plays a role at the level of an individual, but also at the level of a collection of people. In a societal context feedback, and its analysis, always experiences a serious problem: what information is actually measured? Many social variables of interest are only expressed in a qualitative and often in purely subjective way (Barbera and Albertos 1994). This makes it difficult to compare and understand social behavior: even when seemingly identical information is available, the (feedback) action may be completely different. Some important variables are easily measured, like age, salary, unemployment rate. In those instances, the measuring unit is clear, and it is relatively easy to obtain reliable data. But concepts such as, what is aesthetically pleasing, what is humor, attitude, sadness or even concentration are subjective and only understood at a qualitative level. In some cases we may be able to order a concept like good, better and best to compare different observations, but in other instances even that may be problematic, like an appreciation of food.

In a social system it is common to have a mixture of signals, some quantitative, some qualitative. This does not preclude a modeling approach. Relationships between the variables can lead to a valid understanding of the behavior of the system of interest. Models in this case will need to relate qualitative and quantitative signals and most likely will themselves be qualitative in nature. This does not invalidate their usefulness. After all, we all construct such models (somewhere in our brains) all the time.

Let us consider an example of human-with-environment interaction from a control system point of view.

From a control engineering stand-point, the human-environment interaction has all the ingredients that make a closed-loop system. There is sensing, communication, controlled behavior, decision processes, goals to achieve, disturbances to be rejected and so on. Can the systems engineering formalism be used as an advantage to probe this system? With Wiener (1961), we will accept on face value that cybernetics or systems engineering indeed provides a framework that can be used, and perhaps should be used, in the psychological analysis of the human-environment system.

The comparison between the behavior in a human-environment interaction system and the behavior of an engineered control systems will allow one to

- better understand the cognitive processes and their implications,
- develop new theoretical ideas and formulate new hypotheses about psychological processes based on analogy with the easier understood, simpler engineered systems,
- apply control strategies to this challenging field,
- highlight differences and similarities between different psychological approaches.

In the human-environment system, purposeful human behavior requires the formulation of goals, the implementation of actions to reach them, and the ability to evaluate the attainment of these goals. There is an obvious parallel with a control system (Barbera and Albertos 1996). The whole point of a control system is to achieve a control objective (regulate glucose, produce a ceramic tile, point an antenna or deliver irrigation water). There must be sensors that observe whether or not the control objective has been realized. At the core of the control system is a control law or algorithm to compute from the observations which actions the actuators have to take in order to move towards the realization of the control objective. That the analogy is complete must not come as a surprise. Indeed we may well argue that it is our own experience that has dictated the structure of our engineered control loops.

In human behavior there is often a multiplicity of goals, often conflicting goals that guide the behavior at any one point in time. The goals are dynamic, and influenced by the environment. Through our senses, our cognitive processes distill information about both our own actions and internal processes, and the environment. We constantly form some model that enables us to decide on the next actions to take. The model is adapted as new information comes available, and our intentions and our implemented actions are adjusted accordingly. This is really a complex system.

Intentions of course do not always lead to actions. An intention is a necessary but not a sufficient condition to act (Kuhl and Beckman 1992). This level of complexity and sophistication is not often found in engineered control systems, or biological homeostasis. Nevertheless, as control system complexity increases the control system is organized in a hierarchical way. Under such circumstances it is not inconceivable that a local control output, which would normally result in the next actuator action, is not implemented and instead a higher level controller switches in a different command. This is very much like the intention-action model commonly understood in human behavior. As intentions are generated by the model, their enactment is mediated by

- personal perception on how the action may lead to personal goal attainment,
- personal perception on how the action will be judged by the environment, and contribute to the goals valued by the community,
- personal value judgment, which allows one to construct a partial ordering of all intentions and possible alternative actions (Barbera and Albertos 1996).

Actions are neither instantaneously, nor univocally determined by stimuli. Moreover, even under similar external conditions, actions will show great variability across a population of individuals, as each individual will bring a totally different world model to the present observations. This world model is determined by the total past experience of this individual as well as its genetic (memory) make-up.

The idea that emotions are fundamental to any motivational analysis is shared by many psychologists (Zajonc 1984). Emotions are an integral part of our own world model. They are the result of (our personal) processing of the past, comprising all the external inputs, our internal reasoning, all our performed actions, and so on. In this sense our present emotions are seen to be a part of our internal *state*.

The similarities between psychological variables and those in a basic control loop are shown in Fig. 3.22.

These psychological variables are difficult to quantify. Accordingly there is much research and development effort going into trying to correlate these variables with neuronal levels of activity, using for example functional magnetic resonance imaging techniques. Other numerical models are based on carefully crafted tests. Notwithstanding, these variables are often more accurately and more sensibly captured using language, or perhaps using ideas from probability theory or fuzzy set theory (as a tool for approximated knowledge representation). These approaches provide a suitable methodology to capture subjective meaning, which is quite difficult to achieve with purely quantitative procedures.

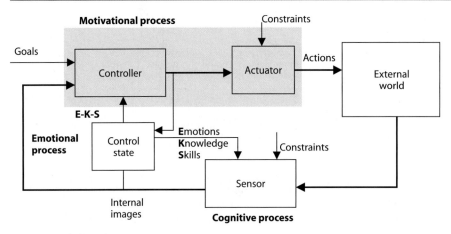

Fig. 3.22. Psychological processes

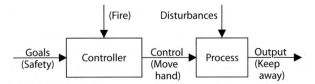

Fig. 3.23. Open-loop control

3.7.1 Simple Control Structures

Let us consider some basic control structures and see how they are relevant in motivational processes (Barbera and Albertos 1996).

Open-loop control. A system without feedback operates in open loop. Figure 3.23 illustrates this.

Some psychological processes referred to as involuntary processes, or reflexes fall into this category. Examples include the celebrated experiments by Pavlov[11], or the reflex we execute when our hand comes in contact with fire or extreme heat (variables shown in brackets in Fig. 3.23). In this case, our actions are not mediated by volition and they are highly repeatable. The trigger for the reflex may be based on feedback, but the action itself is played out without the intervention of further feedback, and the reflex action occurs in open loop.

Closed-loop control. Some processes that are considered as instinctive, but of a more elaborated nature than a pure reflex, are of this type. For example, when hungry, a baby searches for its mother's breast to feed. As the initial objective is achieved, and milk starts to flow, the baby's goal oriented behavior changes to obtaining enough milk. Its

---

[11] Ivan Pavlov, Russian Nobel Prize winner, 1849–1936, famous for his study of conditioned reflexes. His work prepared the way for the science of behavior.

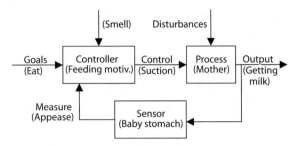

Fig. 3.24. Closed-loop control

suction action will decrease as the hunger is stilled. The feedback in the process is obvious, suction intensity depends on a direct comparison between the goal and the obtained result. Once satisfied, food loses its priority and immediate value and the baby's behavior focuses on something else.

The generic structure is shown in Fig. 3.24. The sensor block in the feedback path executes a relatively complex cognitive process that from the sensory inputs (touch, smell, taste) and responses from the digestive system determines the baby's need for further feeding. This is then translated into motor control to obtain the necessary milk.

Most motivational processes have this closed-loop structure. In particular all those involving volition. The basic characteristic is that action is determined by a comparison between the desired goal and the present state, as determined through our senses.

In applied psychology, it is well-known that feedback plays an important role in successful dieting. The mere goal or need to lose weight rarely results in successful weight loss. Clearly first the need must be established, in that the person must be convinced that there is a need to loose weight. The drive to maintain the diet and its success over time however depends mainly on continuous feedback about progress. As each person reacts differently, feedback must take on different forms. For some peer pressure and regular public *weighing* sessions are essential. Others only need regular self-feedback.

Feedforward control. When disturbances can be measured, predicted or effectively estimated, control can take pre-emptive action to counteract them. Much of our behavior is based on our ability to foresee circumstances. For example, driving a car is almost entirely achieved through feedforward control with little feedback actions required.

In applied psychology effective treatments to give up smoking or other addictions, feed forward plays an important role. According to the Action Control Theory (Kuhl and Beckman 1992), it is important to estimate future disturbances and act on these, either to eliminate these circumstances, or to provide strategies to deal with these. For example it is much harder to give up smoking when constantly exposed to a smoking environment. In such circumstances, effective self regulatory strategies have to be developed to reach the goal of quitting. The use of medication to keep the addiction in check may be essential. Feedback plays an important role too. People feel rewarded and maintain their focus when they have successfully dealt with a difficult situation, like having a meal with close, smoking friends. It is important to prepare for such

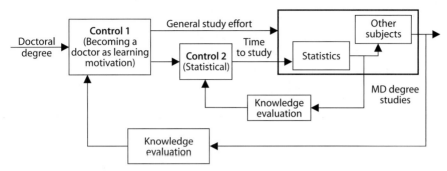

Fig. 3.25. Cascade control: Getting the doctor of medicine (MD) degree

situations and schedule them carefully if at all feasible. These are some of the feedforward strategies that can be put in place.

Cascade Control. Another typical control structure is the use of cascaded control loops, such as encountered in the antenna tracking problem, where four cascaded loops are used to achieve acceptable antenna servo behavior. In general, an inner control loop regulates an internal variable in order to simplify the control actions to be taken in the outer loop, where the actual control objective is achieved. Feedback simplifies the overall behavior, and this is exploited in the cascade structure. The idea is illustrated in Fig. 3.25, where it models the behavior of achieving a medical degree.

The ultimate goal is to become a medical practitioner and to meet this goal one must complete the university studies leading to the bachelor of medicine. The curriculum prescribes a subject in statistics. This results in a sub-behavior in which the required statistical knowledge is assimilated as a matter of completing the medical training. For many, the entire motivation to complete the subject is completely divorced from the subject matter, and entirely driven by the end goal.

As before, the control loop blocks in Fig. 3.25 refer to cognitive processes.

Selective or hybrid control. When control must pursue multiple and perhaps inconsistent goals, selection or hybrid control plays a role. The word hybrid here refers to the mixture of logic or discrete valued signals (like yes or no, or true and false) and analogue signals. As long as goals do not conflict, a single objective that combines all the goals in some meaningful way can be pursued. When goals do conflict we may resolve this conflict by making a choice, or by prioritizing the goals. For instance, we may be motivated to be very fit and exercise in the gym to achieve this purpose. However fitness and gym exercises become irrelevant when the doctor prescribes complete rest because our body must overcome a flu. The control structure involves a new process, as part of its overall control law, namely the process of choice. The point is illustrated in Fig. 3.26.

In Fig. 3.26 the blocks, apart from the process, represent cognitive processes. In general, the goals have a priority that is defined a priori (although we can imagine

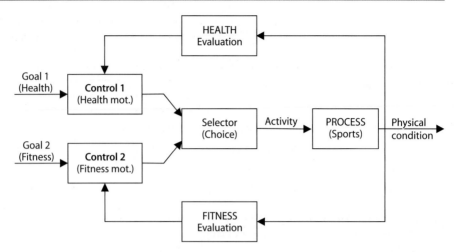

Fig. 3.26. Physical exercise motivation

further cognitive blocks providing the function of prioritization based on past experience). The selector block acts like a guard, deciding which of the alternative strategies have to be followed.

3.7.2 Other Control Approaches

Many concepts in control systems theory have a clear counterpart or interpretation in psychological processes. It is obvious that those control approaches that go under the generic umbrella of intelligent control systems (control systems conceived by the use of artificial intelligence techniques) are closely related to human behavior, as they are indeed inspired by our remarkable cognitive abilities.

Learning control. When there is no clearly defined target, or objective, we must learn what our options are and explore the environment to build up experience. From this experience we may decide where to explore further or perhaps deduce a valid objective. There are special control theories devoted to such schemes, like identification for control (Hjalmarsson et al. 1996; Gevers 1997; Lee et al. 1993).

How we learn to play sports provide an example of such control strategy. At first we are rather tentative and try to improve the required motor skills in some basic way. As our motor ability improves, we develop strategies for co-operative or competitive game play.

In psychology the concept of intention reinforcement plays an important role. This idea is very much related to control schemes such as adaptive control systems (Mareels and Polderman 1994; Astrom and Wittenmark 1988).

Multicriteria and hierarchical control. Human behavior is often motivated to attain a wide range of different goals simultaneously. One particular behavior hypothesis is that people follow after an infinity of different goals (Hyland 1989).

As observed before, in the case of conflicting objectives, choice or priority breaks the dead lock. When there are multiple reinforcing objectives to be pursued they are often hierarchically organized.

Where multiple objectives appear at the same level of importance, we try to strike a balance between the various goals. In optimal control this is known as multi-criteria optimization. A classical example is the natural time-based conflict that arises when someone tries to be a top professional and maintain meaningful family relationships at the same time. Assuming that we indeed do not make an either/or choice between professional recognition and family recognition, our cognitive processes will provide us with a status of the partial attainment of either goal. This information is then used at a local level to further improve achievements towards either goal, but at a higher level decisions are made as to which resources and at what time either goal receives priority. The decision process presents itself naturally in a hierarchical way. Hierarchical control appears under many guises. The classical Maslow's (1954) pyramid provides a good overview.

A particular hierarchy developed for the mind considers:

- Level 1. Unmonitored reflexes and automatic pre-programmed pattern generation, the *open-loop* responses.
- Level 2. *Feedback control* where sensing information is minimally processed and used for rapid tasks. This is typical for our motor skills. Writing, reading and walking are good examples.
- Level 3. The *information-processing* level, where past stimuli are integrated in more complex representations and more complex plans are formulated. It is at this level where we anticipate future actions (the feed-forward). Much of our internal models about behavior, our emotions and the various goals we formulate reside at this level.

It is at levels one and two of the control hierarchy that much of the important work in experimental psychology has taken place.

In a different hierarchy four types of processes are identified:

1. a cognitive process that maps circumstances to notions and knowledge,
2. an affective process which changes the circumstance-action mapping that is currently in force,
3. a cognitive learning routine which builds up the series of circumstance-action rules, and
4. an affective learning routine which builds up the repertoire of reactions and rules by which changing circumstances are related to emotions and moods, generating alterations in the circumstance-action pattern.

> **Feedback in Social Systems**
>
> To some extent, in most of our behavior, we act in order to receive feedback and most of our actions are initiated by feedback.
> Life without feedback would be rather boring.

3.8 Comments and Further Reading

It should be clear that the examples we touched upon are but the tip of the iceberg as far as feedback, control and automation goes. The examples simply reflect some of the personal interests and experiences of the authors.

Process control and automation in manufacturing is a very serious research and development topic in its own right, with many important contributions. A popular introduction is Shinskey (1996), especially useful in a fluid process context. Domain knowledge is critically important in making automation work, and this is reflected in the literature. There are specialized books on automation for just about any branch of industry. Industries that have had a major impact on automation are the automotive, aerospace and defence related industries. Steel (e.g. Kawaguchi and Ueyama 1989) mining and petrochemical industries equally have had major developments as do the utilities, like power and telecommunication. The latter are well-known as examples of *large scale systems* (Siljak 1991), at least from an engineering perspective. The internet, and computing systems more generally are perhaps the first examples of engineered systems that approach biological systems in terms of their complexity. As the technology to interconnect almost any sensor and actuator in our engineered environment into an internet-like communication network becomes ubiquitous the complexity of control systems will increase dramatically, but so will the capacity to manage our environment.

The theory and practice of adaptive systems (Astrom and Wittenmark 1988; Mareels and Polderman 1994; Goodwin and Sin 1984; Anderson et al. 1986b) has a rich history in systems engineering, starting from the early work around the so-called MIT rule (Whitaker 1959) for adaptive control. Learning control, iterative control, extremum seeking control are all variants starting from this simple idea of a self-optimizing control law.

The automata and finite state machines as well as their extensions in event-based systems have equally a long history with many important contributions (Cassandras and Lafortune 2008). This whole line of work builds on the contributions by Von Neumann and Turing, who clearly saw their work in the light of cybernetics, see for example Turing (1992) and Von Neumann (1958).

Equally our discussion of homeostasis is but scratching the surface of where feedback and systems engineering play a role or are important to understand biological processes. Mathematical biology has a rich history, and the developments in systems biology are just too explosive to be able to do them justice here. For some of the earlier work where feedback clearly takes center stage, see for example Mees (1991) and Goldbeter (1997).

The use of systems engineering ideas in social systems and psychology is a natural extension of the early cybernetics ideas. The origin of which may be found in the MACY conferences, held between 1946 and 1953 (see, for instance, the American Society of Cybernetics: *http://www.asc-cybernetics.org/*). For a contribution discussing cybernetics at the level of the behavior of a society, by Norbert Wiener himself, see for example Wiener (1954).

Chapter 4 Signal Analysis

4.1 Introduction and Motivation
4.2 Signals and Signal Classes
4.3 Signal Transforms
4.4 Measuring a Signal
4.5 Signal Processing
4.6 Recording and Reproduction
4.7 Comments and Further Reading

Chapter 4
Signal Analysis

Divide et impera,
Divide and conquer

4.1 Introduction and Motivation

So far we have seen that signals provide information about systems. Some systems (measurement devices) measure signals and all systems generate signals. Signals and systems are essentially inseparable. This chapter focuses on signals.

All signals are obtained from observing over time the physical world around us using a sensor or measurement system. Sensors extend the capability of our own body senses: vision, sound, touch, smell and taste. Moreover, we shall see that signals can be represented with an advantage using a network of simple systems.

In the picture introducing this chapter, the discrete footprints or the continuous bike trail are both providing position information. Tracking along the trail with a GPS[1] receiver we can associate with the trail a time sequence of absolute positions on the globe. In this fashion the trail becomes a signal, a function of time. A different signal derived from the trail could be to observe our velocity and heading as we are following the trail.

Questions like, what is a signal, how do we obtain a signal, how do we communicate a signal, what properties does a signal have and how do we distinguish signals are briefly discussed in this chapter. These questions belong to the realm of *signal processing*.

The same signal may have more than one mathematical or physical representation that is equivalent from the point of view that different representations carry the same information. Depending on the circumstances one representation can be much more instructive or advantageous than another. For example one representation may require much less storage space on a computer than another, and economy of representation is important in applications. In particular networks of (simple) systems can be used to construct signals.

Fundamental to the study of signals and systems is the idea of *divide and conquer*: a signal is considered as a combination of much simpler, easier to understand, signals; or as an output from a network of (simple) systems.

The notion of periodicity, the fact that a signal repeats itself after a finite amount of time, which is called the period of the signal, is an extremely simplifying notion. Periodic (or nearly periodic) signals are ubiquitous in nature. Perhaps remarkable but almost any signal can be considered as an appropriate linear combination of sinusoids, which form a class of very well behaved periodic signals. Understanding how a signal can be decomposed into a sum of sinusoids is the object of *spectral* or *Fourier* analysis.

---

[1] Global Position System.

Although the basic ideas of spectral analysis were developed in the early 19th century by Fourier[2] for the purpose of understanding heat and heat transfer, the main applications are much more recent. Indeed spectral analysis underpins many of the significant achievements in compactly representing signals in binary formats; such as

- MP3, the standard for audio file compression developed by the Moving Picture Experts Group or MPEG as part of the MPEG-1 standard used to encode video and audio information onto video compact disks, and
- JPEG, Joint Photographic Experts Group, a subcommittee of the International Standard Organisation, charged with the development of compression and digital storage standards for image information, and
- MPEG, a standard for pictures, video and audio information.

Furthermore, spectral analysis plays a critically important role in such diverse applications as RADAR (RAdio And Ranging), SONAR (SOund NAvigation and Ranging) and radio-astronomy. Perhaps more significant without it there would be no such thing as a mobile phone or wireless internet.

The chapter starts with an overview of how we may represent a signal, followed how we may classify signals according to various properties. The main idea of signal processing, or how to extract information from a signal is briefly introduced.

4.2 Signals and Signal Classes

As discussed before, signals are functions of time. The collection of all functions of time is enormous, and one of the main aspects of signal processing is to bring some order and structure to this unwieldy large collection.

The following basic information which we need to supply in order to define and understand a signal[3], also helps us to distinguish various classes of signals.

- The *time interval*, called the *time domain* over which the signal is defined. It can be finite, or infinite.
- The way we measure time. Time instances can form a *discrete* collection of points on a real line, or the time instances can form a *continuous* interval on the real line. In the discrete time case, the consecutive time instances can be regularly spaced in that the difference between successive time instances is constant, or irregular. The latter has distinct advantages that are exploited to measure periodic signals with great precision.
- The (numerical) values the signal can take, the signal's *range* as well as its units; discrete, or continuous; finitely many or infinitely many values.

---

[2] Fourier, Jean Baptiste Joseph, (March 21, 1768–May 16, 1830), French mathematician famous for his study of heat transfer which was published in 1822. He was the first to use sums of sinusoids to approximate general signals.

[3] In practice, how a signal is measured is even more important. Which sensor has been used? How was the sensor calibrated? Without this vital information a signal is virtually meaningless. It is tacitly assumed that this information is available.

In practice as we can only resolve time with finite precision, over a finite interval of time, and the transducers can only resolve the signal value with finite precision, all measured signals are defined on a finite set of discrete times and only take finitely many values. Both domain and range are finite, discrete sets: the value of a signal and the time instant it happens, if recorded in digital form, will be represented by a finite number of bits. Unavoidably this introduces errors and information degeneration. Nevertheless, the abstraction of considering signals as defined on an interval of the real line and that take on values in a continuum enables powerful analysis tools not available for discrete valued signals. This raises the question: how do we associate a signal that has a continuous domain and range to an observed signal that has a discrete domain and range? Models and systems play an important role in answering this question.

4.2.1 Signals Defined through Mathematical Algorithms

Signals that have a precise mathematical representation in that we are able to compute the signal value *reliably* are called *deterministic*. Some of them are depicted in Fig. 4.1.

Constant Signal

A constant signal is defined as $s(t) = a$, constant, independent of the argument t. It is a rather boring signal. For example the speed of light is a constant signal. As it does not change with time, its derivative (with respect to time) is null. This can be expressed by

$$\frac{ds(t)}{dt} = \dot{s}(t) = 0 \tag{4.1}$$

Alternatively

$$s(t) = s + \int_0^t 0 \, d\tau \tag{4.2}$$

In system language, we may interpret the above equations as stating that constant signals are modeled as the output of an integrator (which is a system) with zero input and initial value equal to the constant.

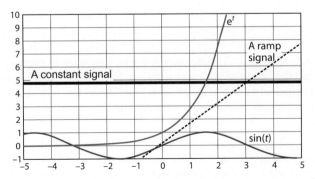

Fig. 4.1. Some simple signals represented over a finite interval of time, as continuous valued signals

Polynomial Signals

A polynomial signal is a finite linear combination of time raised to an integer valued power $t, t^2, t^3, \ldots$, including $t^0 = 1$. It is of the form $s(t) = a_0 + a_1 t + \ldots + a^{n-1} t^{n-1} + a_n t^n$ for scalars a_i $i = 0, 1, \ldots, n$. Such polynomials are said to be of degree n, if $a_n \neq 0$, as this is the highest power of time appearing in the expression.

For example the signal $s(t) = 1 + t - t^2 + t^3$ is a cubic polynomial signal, or a polynomial of degree 3.

A constant signal, $s(t) = a$ where a is a scalar, independent of time, is a special case of a polynomial signal. The importance of polynomials may be appreciated from the fact that any signal (on some finite interval) can be arbitrarily well approximated by a polynomial signal. This is a useful property, for instance, to interpolate or extrapolate signals from its samples. This result is due to Weierstrass[4].

A *ramp* signal is a first order polynomial signal of the form $s(t) = at + b$ for some scalars a, b. A ramp signal can represent the position of a photon along a glass fiber in a telecommunication network or the shaft angular position under constant angular speed. The derivative of a ramp signal is a constant signal. In fact, if $s(t) = at + b$, its derivative is

$$\frac{ds(t)}{dt} = \dot{s}(t) = a$$

Quadratic polynomials, that is signals of the form $s(t) = at^2 + bt + c$ for some scalars a, b, c are useful to represent the trajectory of a golf ball, or the trajectory of a ballistic missile such as an arrow. Again, its derivative is a ramp signal and the second derivative is a constant signal

$$\frac{ds(t)}{dt} = \dot{s}(t) = 2at + b$$

$$\frac{d\frac{ds(t)}{dt}}{dt} = \frac{d^2 s(t)}{dt^2} = \ddot{s}(t) = 2a$$

A simple network of systems that computes any polynomial of degree 3 or less is represented in Fig. 4.2. The network consists of only two different systems, adders and multipliers, arranged in a linear fashion from left to right, alternating an adder followed by a multiplier. The multipliers, apart from the first multiplier on the left, receive two inputs, namely the time instant at which we want to evaluate the polynomial signal and the output of the adder to its left. The first multiplier receives as input the leading coefficient as well as the time instant. The adders receive also two inputs, a coefficient of the polynomial to be evaluated and the output of the multiplier to its

---

[4] The result is known as the Weierstrass approximation result. Karl Theodor Wilhelm Weierstrass, 1815–1897 was a German mathematician, best known for his work on complex function theory. He is one of 300 mathematicians who had a lunar feature named after him: the Weierstrass Crater.

left. It is easy to see how to construct a network to compute all polynomials of any degree, which will contain as many multipliers and adders as the degree of the polynomial we want to compute.

An alternative format to obtain polynomial signals is to use a cascade, or series connection of integrators with a constant input, and each integrator has some initial value. See also Fig. 4.3. The advantage over the previous system is that here the inputs are constants.

Exponential and Sinusoidal Signals

Exponential signals are of the form $s(t) = be^{at}$, where a, b are scalars and e is a positive scalar[5] It is easy to realize that as time progresses that the signal value will grow if $u > 0$ and it will decrease to zero if $a < 0$, although it will take and infinite amount of time to vanish. This evolution is referred to as exponential decay to zero.

Exponential signals can be defined as those signals that are proportional to their derivative:

$$s(t) = be^{at} \quad \text{if and only if} \quad \dot{s}(t) = as(t)$$

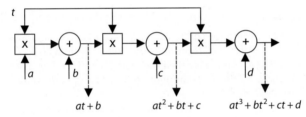

Fig. 4.2. A network of systems that computes arbitrary polynomial signals up to degree three. The external inputs to the network are the signal t, and the constant signals a, b, c, d the coefficients of the polynomial signal. The network computes a number of different polynomial signals of degree 1 to 3. Identified as outputs of the network

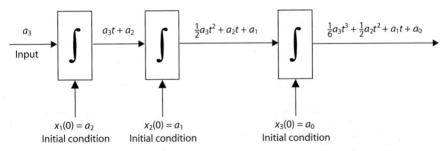

Fig. 4.3. A cascade of three integrators that computes arbitrary polynomial signals up to degree three

---

[5] e is approximately 2.718281828 and sometimes described as Euler's number. Euler, Leonhard 1707–1783, German mathematician, one of the most famous mathematicians in history with contributions across many different fields.

This property will be very useful in analyzing signals because anytime we encounter an equation of the form

$$\dot{y}(t) = ay(t) \tag{4.3}$$

then it means that $y(t)$ is exponential, that is, the solution of Eq. 4.3 is an exponential function.

Signals of the form $s(t) = h\sin(2\pi t/T + \phi)$ are called sinusoidal signals. We have already seen these kinds of oscillating signals in Fig. 2.12, when dealing with a frictionless mass and spring system. The signal is defined by three parameters: the amplitude, h, the period, T, and the phase, ϕ. This a fundamental class of periodic signals.

Exponential, sinusoidal signals, a product of an exponential and a sinusoidal signal and linear combinations of such signals are of particular importance in the study of linear systems.

Also, exponential and sinusoidal signals are closely related. If in e^{at} the scalar a is *complex* $a = j\omega$, with ω real and j the symbol to denote $\sqrt{-1}$ then the signal is actually a complex sinusoidal signal, namely $e^{j\omega t} = \cos(\omega t) + j\sin(\omega t)$ (this relationship is due to Euler).

Complex Numbers

A complex number is composed of two parts, a real part and an imaginary part, usually written as $z = a + jb$. Here j represents the unit in the imaginary component. It has the property that $j^2 = -1$. a is called the real part and b is the imaginary part of the complex number z. Complex numbers can be seen to correspond to points in the plane.

A complex number can be represented by its modulus and argument, defined as

$$r = \sqrt{a^2 + b^2}\ ;\ \ \beta = \arctan\frac{b}{a}$$

respectively. Introducing,

$$r = e^{\alpha}\ ;\ \ e^{j\beta} = \cos(\beta) + j\sin(\beta)$$

any complex number can be written as

$$a + jb = e^{\alpha + j\beta} \tag{4.4}$$

To add two complex numbers, add the real and imaginary parts as:

$$(a + jb) + (c + jd) = (a + c) + j(b + d)$$

To multiply two complex numbers, use the following recipe:

$$(a + jb)(c + jd) = (ac - bd) + j(bc + ad)$$

from which it follows that $j^2 = -1$.

From the above, a rather remarkable property can be deduced:

$$e^{j\pi} = -1$$

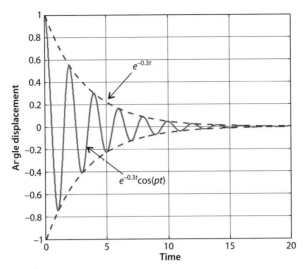

Fig. 4.4. The angle of a swinging pendulum subject to friction

Radioactive decay for example is modeled using exponential signals, as in fact the change in radioactive material is proportional to the amount of available material. Thus, it can be modeled as in the Eq. 4.3.

The swinging arm of a pendulum is modeled using sinusoidal signals. If there was no friction in the pendulum a purely sinusoidal signal would be observed. In the presence of friction the response would be a sinusoid whose amplitude is decaying like an exponential. This is illustrated in Fig. 4.4, showing a clear analogy with the damped mass and spring evolution shown in Fig. 2.13.

Sinusoids form a class of time functions particularly well suited to describe periodic phenomena. In order to illustrate what a sinusoid is and how it is closely related to linear systems we elaborate a little. Consider a clock with a single hand and assume that the hand rotates at a constant angular speed that is the angle the hand sweeps through is proportional with lapsed time. The length of the orthogonal projection of the hand onto any diameter of the circle traced by the hand defines a sinusoidal signal. A similar arrangement is found in a slider-crank mechanism illustrated in Fig. 4.5 that converts rotary motion into a horizontal back and forth motion[6]. Clearly the linear displacement signal is periodic, because the position of the hand is a periodic function of time and also of angular position[7].

---

[6] If the conversion is considered the other way around, a back and forth motion can be converted into a rotary motion, in this form it is applied in steam locomotives.

[7] Newtonian mechanics gives us a model for uniform rotary motion, if $r = (x, y)$ represents the position of the hand, with mass m, with respect to the fixed point around which it rotates, then the following balance equality holds: acceleration times mass + centripetal force = 0. Mathematically, it is $m\ddot{r} + m\omega^2 \bar{r} = 0$, or deleting the mass, we get the dynamic motion equation $\ddot{r} + \omega^2 \bar{r} = 0$. The projection of this position onto the horizontal satisfies $\ddot{x} + \omega^2 x = 0$. The solutions of this equations are called the *sine* functions.

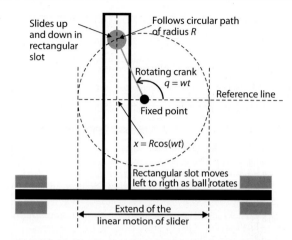

Fig. 4.5. A slider-crank mechanism converts a rotary motion into a periodic linear motion. A rotary motion with constant angular velocity leads to a sinusoidal linear motion

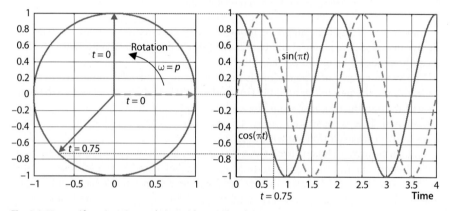

Fig. 4.6. How uniform rotation and sinusoids are related

Usually a sinusoidal continuous time signal is represented as

$$s(t) = h\sin(\omega t + \phi) \tag{4.5}$$

The word "sin" is the mathematical abbreviation for the *sine* function. t represents time, measured in seconds and the angle ϕ, measured in radians[8], is known as the initial *phase* of the signal, and S_0 is the maximum value (magnitude) of the signal. The phase

---

[8] Radian is a unit for angle measurement and there are 2π radians in a full circle. Essentially an angle is measured by its arc length on the circumference of a circle with unit length radius. π is a transcendental number, it is the ratio of the circumference of a circle and its diameter, approximately given by $\pi = 3.1415926535$.

determines the time reference. ω is known as the pulsation of the sinusoid, being expressed in rad/s (radians per second). h is known as the amplitude.

$T = 2\pi/\omega$ is the period. The frequency is the number of periods per unit of time, $\nu = 1/T$. The unit of frequency is Hertz[9], abbreviated as $1\,\text{Hz} = \text{s}^{-1}$. When $\phi = -\pi/2$ the function is called cosine and denoted as $\cos(\omega t) = \sin(\omega t - \pi/2)$.

4.2.2 Periodic Signals

The swinging arm of a grandfather clock or metronome (Fig. 4.7) ticks periodically. By design it measures equal time intervals between the extreme positions of the swinging pendulum. By way of illustration, the angle of the pendulum is graphed against time in Fig. 4.8. Obviously there is a part of the picture that repeats itself; we say that the angle is a periodic signal.

A signal s is called *periodic*, provided there is a time interval T such that shifting the signal over T units of time leaves the signal unchanged, that is $s(t) = s(t + T)$ for all time t. The smallest time interval with this property is called the *period*.

Fig. 4.7. A metronome is used to provide a reference of time

---

[9] The unit is named in honor of Heinrich Rudolf Hertz, who was born on February 22, 1857, in Hamburg, Germany and died January 1, 1894, in Bonn, Germany. He studied engineering and became a Professor of Physics at Karlsruhe University at the age of 28. In 1885 he succeeded in the first practical demonstration of electromagnetic waves that were predicted by Maxwell's theory of electromagnetism, which paved to road to wireless communication.

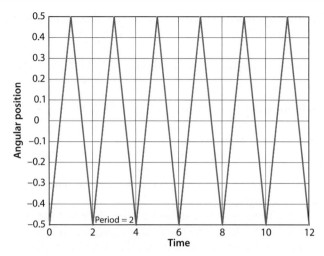

Fig. 4.8. A metronome's angle plotted against time or a so-called (periodic) saw-tooth function

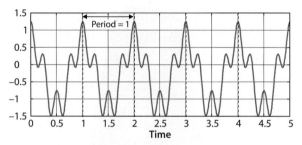

Fig. 4.9. A periodic signal, with period 1

Periodic phenomena are quite common, and in our modern society we have chosen to organize our lives according to periodic phenomena. We are used to yearly, seasonal and daily patterns. In biology, heart rhythms, breathing rhythms and hormone cycles play an important role. They are not really periodic but nearly so (in fact periodicity normally indicates that something is wrong). In engineering systems, periodic motion, because of its simplicity, is everywhere. The shaft of an electrical motor or generator, the position of a cylinder in an internal combustion engine, the angle position in a gearbox, the read-position of a compact disk player or tape-recorder are all characterized by periodic signals. Also the tones produced by musical instruments, traffic light and train schedules, all of these are periodic signals or can be modeled as periodic signals.

A portion of a particular periodic function is illustrated in Fig. 4.9, in principle the time axis has to be extended indefinitely (both forward and backwards in time), but of course we cannot represent that in a picture.

A periodic signal is deterministic and predictable. Indeed it suffices to observe a periodic signal over a single period in time (it does not matter which period is observed) to predict its value for any instant (in the future) or reconstruct its value for any instant in the past. All periodic signals are deterministic, but not all deterministic signals are periodic.

Moreover almost all periodic signals of period T can be represented as a sum of sinusoidal signals (Eq. 4.5):

$$\sum_{k=1}^{k=\infty} h_k \sin\left(\frac{2k\pi}{T}t + \phi_k\right) \tag{4.6}$$

$1/T$ is called the *fundamental* frequency, the multiples being denoted as *harmonics*. Such expressions are called *Fourier series*. The concept is illustrated in Fig. 4.10, where a discontinuous wave of period 2π is successively approximated by its fundamental and the first 3 harmonics. Notice that the approximants are continuous periodic functions. Hence we have to be careful in which way we interpret that the approximants are indeed close to the function of interest and get closer as we include more harmonics[10].

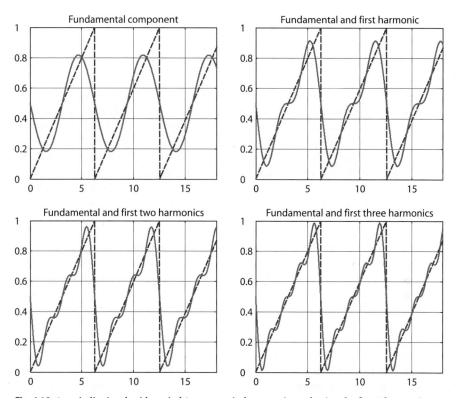

Fig. 4.10. A periodic signal, with period 2π, successively approximated using the first 3 harmonics

---

[10] A usual measurement of the goodness of the approximation between two periodic signals is to evaluate the total error or difference between signals over one period. The cost or error index can be expressed in different ways, the integral of the square error along a period being the most popular one. In the next chapter, the Least Square index will be used to measure the approximation degree (the "fitness") in generating arbitrary signals.

Appropriate sums of sinusoidal signals can be used to approximate arbitrary signals. This class of signals has enormous flexibility. For almost any signal $s(t)$ it is possible to find an appropriate collection of amplitudes and frequencies and phases to approximate $s(t)$ as a sum of sinusoids. Observe that although each term in such sum is a periodic signal the sum is not necessarily periodic. If there are finitely many terms in the sum, we say that the signal is quasi-periodic. In general the signal represented as a sum of sines, with arbitrary pulsations, is not periodic, being also referred to as *aperiodic*.

4.2.3 Stochastic Signals

So far we discussed signals that can be represented as sums of sinusoidal signals. All such signals are deterministic, they are entirely prescribed.

Nevertheless as soon as a signal is measured, or indeed generated in some way it will contain some unpredictable aspects that are rather a property of the way the measurement was taken than of the signal itself.

The process of measurement required to observe any signal typically involves some irreproducible effects (that are being minimized of course to ensure that measurement represents the signal of interest with great fidelity) due to the measurement technique. To capture this notion of irreproducible effects, the term *stochastic* or *random* is used.

In describing a random signal, it is more important to understand the ensemble of possibilities than the particular realization of the signal. A typical example of a highly unpredictable, random signal is the outcome of a series of dice throws. At each instant of time the outcome of the dice is a *random variable* that takes on integer values between 1 and 6. If the dice are so-called fair, all of the outcomes are equally likely. So, in a long sequence of throws we may expect to see an equal number of occurrences for all the possible outcomes of a dice throw. The reason why the throw of a dice is considered random is because we are not able to precisely control the experimental conditions of the throw. Every time we throw the dice the conditions are slightly different, resulting in a different outcome. However if a machine was designed to throw the dice in a reproducible way, it would create a very predictable and actually a constant outcome for each throw of the dice. A single observation of the result suffices to predict the signal forever.

Just as we cannot predict the outcome of a dice throw, in most measuring devices the precise conditions of taking a measurement are also never entirely reproducible, or precisely known, and to describe this we use the notion of random signals. The calibration of an instrument is about understanding and characterizing the ensemble of possibilities of the map from signal value to measurement value.

Random signals are also used to represent things we do not know, or cannot possibly know. For example in the process of sampling the air pressure impact on a microphone and representing it as an integer for recording onto a compact disk involves by necessity a selection of an integer nearest to the (scaled) measurement. Some form of rounding error is introduced. The outcome of the rounding process is quite deterministic; given a reading of the air pressure by the transducer there is precisely one integer that captures this reading best. However given the integer representation, it is not clear at all which air pressure this actually came from (there is a whole range of air pressures that could have lead to the same integer). To capture this lack of knowledge, we often *model* the actual air pressure as the measurement (known) plus a random component that represents the unknown rounding error. Such a representation is very useful when comparing different signals, for example when discussing the sound fidelity of different recording techniques. Is analogue better than digital? The answer lies in the nature of these random components (there are random components in both recording techniques).

Randomness also allows us to define the behavior of signals not precisely known. For instance, if we consider the position of a mechanical shaft, some internal vibrations may influence its position in a complex way we do not want (or we are not able) to model. Then the position is considered to be disturbed by a random signal.

4.2.4 Chaotic Signals

The emphasis on *reliable* computation in defining a deterministic signal is important. The mere availability of a recipe or algorithm to compute a signal may not be sufficient nor practical for its actual computation.

For example, the recipe $s(n+1) = 4 \cdot s(n) \cdot (1 - s(n))$ for $n = 1, 2, \ldots$ with a starting value $s(1)$ in the interval $(0, 1)$ appears reasonable enough. Yet, due to the nature of the recipe and because computers have only a finite collection of numbers to work with, the actual recipe cannot be used to compute the signal over an arbitrarily long interval. Indeed, in order to compute $s(n)$ with a precision of N bits requires one to specify $s(1)$ to $N + n - 1$ bits[11]. For large n this is not a practical proposition, as the required computing resources grow linearly with n. The sequence $s(n)$ is characterized as *chaotic*. (See Fig. 4.11.)

We can use this loss of precision as time progresses to our advantage (every coin has two sides). Indeed, if we had known that the signal s obeys this relationship, and we measure a sequence $s(1)$ to $s(n)$ with an instrument that provides us the values of s with x bits of precision, then we are able to determine $s(1)$ to $x + n - 1$ bits of precision.[12]

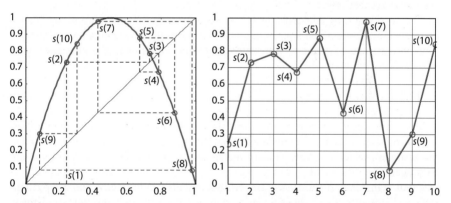

Fig. 4.11. A short sequence of observations of the chaotic signal $s(n+1) = 4 \cdot s(n) \cdot (1 - s(n))$, for $n = 1, \ldots, 10$ with $s(1) = 0.241$

---

[11] Loosely speaking that is, the statement is correct for large n and is generic in the choice of $s(1)$. For example it certainly is not a true statement for $s(1) = 1$ or $s(1) = 0$ or $s(1) = 3/4$ to pick but a few exceptions.

[12] Measuring $s(1)$ and $s(2)$ provides us $2x$ bits of information, but because we know how $s(2)$ is generated from $s(1)$ through the recipe, we can infer 1 bit of additional information about $s(1)$.

A digital computer implementation of this algorithm will not be able to truly generate the chaotic signal we refer to above. Indeed as the computer can only represent finitely many different numbers between 0 and 1, any sequence generated must eventually repeat itself, i.e. the computer generated signal using this algorithm is necessarily periodic due to approximation errors. This phenomenon is known as the *collapse of chaos*.

4.3 Signal Transforms

As indicated the collection of sinusoids may not be the best collection of functions to represent the signals of interest. Different signal environments require their own appropriate collection of *basis* functions. Representing a time function as a linear combination of a collection of basis functions is in general called a *signal transform*.

The Fourier transform, which generalizes the Fourier series (Eq. 4.6) is one such transform. It uses as basis functions the class of sinusoids of any pulsation. Almost any signal $s(t)$ can be represented in a unique manner as a (infinite) linear combination of sinusoids:

$$f(t) = \mathfrak{F}^{-1}\{F\}(t) = \int e^{j2\pi\nu t} F(\nu) d\nu \tag{4.7}$$

$$F(\nu) = \mathfrak{F}\{f\}(\nu) = \int e^{-j2\pi\nu t} f(t) dt \tag{4.8}$$

The Fourier transform $F(\nu)$ is often called the frequency domain representation or the spectral representation of the (time domain) signal $f(t)$. In signal processing, the terms spectrum and bandwidth refer to how important particular pulsations are in its Fourier transform, and this part of the spectrum that contains half the total energy of the signal respectively.

The energy of a signal equals the integral over time of the square of the signal. It is an important theorem[13] in Fourier analysis that states that the energy as measured in time domain, equals the energy measured in frequency domain (the theorem is a lot deeper than it sounds).

$$\int |f(t)|^2 dt = \int |F(\nu)|^2 d\nu \tag{4.9}$$

There are many different types of transforms. Some are more useful than others. The usefulness of the transform depends on the properties of the basis functions and the ease with which one can go from the signal to its transform representation and back again.

Other transform methods that are well documented in literature, and are particularly well suited to the study of linear systems are the Laplace[14] and the z-transform. The Laplace transform is for continuous time signals, and uses as basis functions the class of signals $e^{at}\cos(\omega t)$ and $e^a t \sin(\omega t)$, which are conveniently summarized in com-

---

[13] The theorem is due to Michel Plancherel, Swiss mathematician, 1885–1967.
[14] Pierre-Simon Laplace, 1749–1827, famous French mathematician, well-known for his contribution to statistics and mathematical astronomy.

plex notation as $e^{(a+j\omega)t}$. This transform needs a complex variable to specify the basis functions $s = a + j\omega$. The transform is a complex valued function. There is an inverse transform, taking us back from the complex transform function to the time domain signal. There are mathematical tables that summarize typical functions and their Laplace transforms. Formally, the Laplace transform of the signal $f(t)$ is given through an integral:

$$F(s) = \mathcal{L}\{f(t)\} = \int_0^\infty e^{-st} f(t) dt \tag{4.10}$$

where s is a complex variable. $F(s)$ denotes how important the basis function e^{st} is in the representation of $f(t)$, which is known as the inverse Laplace transform:

$$f(t) = \mathcal{L}^{-1}\{F\}(t) = \frac{1}{2\pi j} \int e^{st} F(s) ds \tag{4.11}$$

For example the following Laplace transforms are easily computed based on the previous definition:

$$\mathcal{L}\{K\} = \frac{K}{s}; \quad \mathcal{L}\{e^{at}\} = \frac{1}{s-a}$$

Among the many properties of this transform we must mention two. First *linearity*: the Laplace transform of a linear combination of signals is the linear combination of their Laplace transforms. Also, the Laplace transform of the derivative of a function is

$$\mathcal{L}\left\{\frac{df(t)}{dt}\right\} = sF(s) - f(0) \tag{4.12}$$

These two properties are useful in representing linear systems in continuous time.

> **Representing Signals As Sums of Sinusoids**
>
> Arbitrary signals measured over some time interval can be approximately represented, to any degree of accuracy, by a sum of sine functions. The collection of frequencies required to represent the signal is called its spectrum. The signal's bandwidth is that interval of frequencies that represent half the signal's total energy.

4.4 Measuring a Signal

Signals are measured through the use of some sensor or measurement system. The signal to be measured is its input, whereas some disturbances may act on it, and the output of the sensor is the measured signal.

Measured signals always differ from what we really intend to measure (see also Fig. 4.12). Indeed more often than not, the very act of observing or measuring signals changes what we expected to observe. Also the sensor may be influenced by other signals that may exist in the environment, besides the desired signals. These signals we typically call *noise* or *disturbances*.

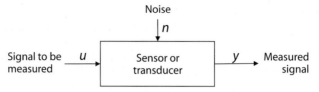

Fig. 4.12. Measuring signals requires a measurement system

Other differences between signal and measurement are due to the limitations of the sensor itself. Neither the resolution in time nor in signal value is arbitrarily small. Also the dynamic response of a sensor is limited. Signals cannot change too fast or the sensor will loose track, neither can the signal be too large, or the sensor will saturate or even fail.

4.4.1 The Size of a Signal

In characterizing a signal as well as studying the effect signals can have, the property of size matters. Signal size can be measured in a variety of ways.

The maximum value of the absolute value of the signal is one measure of size, often used because of its simplicity. The same can be said about the maximum magnitude of the frequency components of the signal. Nevertheless, a simple maximum has the disadvantage that the maximum value may represent a rather exceptional circumstance for the signal; for example a signal that takes on a large value for a very short period of time only. A measure of size that captures more the average of the behavior over time is often more appropriate. For instance, in a chemical plant the transient instantaneous composition may not be that relevant (as long as it is safe) and what matters more is the average composition. Also, when monitoring a signal around a specified value, a small deviation sustained over a long period of time is often more important than a short transient large deviation.

For signals, like sinusoids, a measurement *averaged* over a period may be more appropriate. In order to always get a positive value of the size we can consider the absolute value, $|s(t)|$, or the square of the signal, $s^2(t)$ averaged over one period T such as

$$\|s\|_m := \frac{1}{T}\int_0^T |s(t)|\,dt \qquad (4.13)$$

The following measure is common in power engineering:

$$\|f\|_{rms} := \sqrt{\frac{1}{T}\int_0^T f^2(t)\,dt}$$

being referred to as the *root mean square* value, *rms* for short. For instance, when we say that voltage of the electrical power we use at home is an alternating signal, a sinusoid, of 230 V (Europe), or 120 V (USA) or 240 V (Australia). This number 230, 120 or 240 is precisely the rms value of the domestic voltage wave in that locality. It has the physical meaning of identifying a DC voltage (constant voltage) with the same energy over one period as the actual AC (periodic, sinusoidal) voltage.

4.4.2 Sampling a Signal

The basic idea of *sampling* a signal is presented in Fig. 4.13. In Fig. 4.13a the underlying signal (bold continuous line) is sampled regularly in time. Samples are indicated by the small rectangles. Because there is a minor jitter in the actual time instants that a sample was taken and because the signal value is resolved with a certain finite precision, a sample point only informs us that the actual signal passes somewhere through the small rectangle centred at the sample point. Regular sampling in time is very common. It is used in almost all video and sound recording techniques. For example CD (compact disk) audio is typically recorded with 44 100 samples per second (the difference between sample instants is about 23 micro seconds) with 16 bit resolution, or 65 536 different levels of sound intensity. Depending on the recording technique there may be a single sound channel, so-called mono, or two sound channels, so-called stereo, and more channels (as many as 7) for surround sound effects.

In Fig. 4.13b the underlying signal is sampled regularly in signal value, or range space. This is very common in supervisory control and data acquisition (SCADA) systems. For example if we want to monitor a liquid level in a tank, it is more useful to know when the liquid level reaches a certain level than to know what level the liquid has at a given time (especially if we want to avoid spilling and/or have to maintain a minimum level). In this case the rectangles are wider in the time domain than the range domain, as we need to determine time. There is also a small error in the threshold detection. Typically, for slowly varying signals, regular sampling in range space rather than domain space provides a good representation of the signal with far fewer samples required than regular sampling in time. The advantage and disadvantage of sampling in domain space is that the sampling rate varies with how the signal varies over time. Regular sampling in range is also called *Lebesgue* sampling[15] or *event*-based sampling.

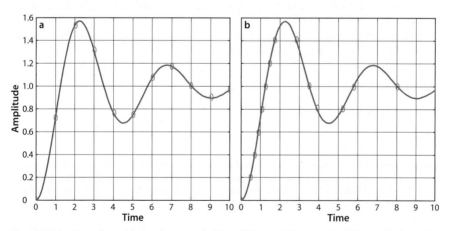

Fig. 4.13. Sampling of continuous time signals (the *solid curve*). The sample values are indicated by *small circles*. **a** Sampling regularly in time. **b** Sampling by signal levels or Lebesgue sampling

---

[15] Henri Lebesgue, 1875–1941, French mathematician, famous for his formulation of integration and measure theory. He is one of the mathematicians with a lunar feature named after him.

4.4.3 Sampling Periodic Signals: the Aliasing Problem

Let us consider what happens if we sample a periodic signal. Assume that the sampling process is ideal in that it measures a signal instantaneously and with infinite precision. Consider equidistant samples are taken, that is the sampling instants are $t_k = kT_s$, for integer k and where $T_s > 0$ is the sampling interval. Sampling the sinusoidal signal (Eq. 4.5), $s(t) = h\sin(\omega t + \phi)$, leads to the measured signal s_s given by

$$s_s(k) = s(kT_s) = h\sin(\omega kT_s + \phi) \quad k = 0, 1, \ldots \tag{4.14}$$

Sampling with a sample period T_s that is rationally related to the period of the signal s leads to a discrete time periodic signal s_s, and there is an integer K such that $s_s(k+K) = s_s(k)$ for all integers k. This is the case when

$$\frac{T_s}{T} = \frac{T_s \omega}{2\pi}$$

is a rational number. The period of the sampled signal in time units is KT_s. It is at least as large as the period of the signal s itself, ($KT_s \geq T$), see for example Fig. 4.14.

For $T_s = T$ we actually obtain that s_s is a constant signal! This phenomenon is called *aliasing*. Aliasing can only be avoided provided $2T_s < T$, or the sampling frequency $f_s = 1/T_s$ must be at least twice the frequency of the signal we want to sample. The sampling illustrated in Fig. 4.14 satisfies this criterion. This fundamental observation is due to Nyquist[16] and Shannon[17]. It applies in great generality when sampling a particular signal (not necessarily a periodic one) with a constant sampling frequency, then this sampling frequency must at least be twice the largest frequency in the signal's spectrum in order to avoid aliasing. The nature of the significant loss of information associated with aliasing is illustrated in Fig. 4.15.

Aliasing can be avoided by ensuring that the signals we are going to measure are sufficiently slow, or by preprocessing the signal before samples are taken and eliminating all the fast signal variations that would lead to aliasing. This requires a so-called *anti-aliasing filter*.

---

[16] Harry Nyquist, born in Sweden February 7, 1889, died in the USA April 4, 1976. He was a telecommunications engineer who made important contributions to information theory and systems theory. He came to the USA in 1907, and obtained a Ph.D. in physics at Yale University in 1917. He worked as a research engineer at the Bell Telephone Laboratories until his retirement in 1954. He is one of the inventors of the fax machine.

[17] Claude Elwood Shannon, 1916–2001, is the founder of information theory with his famous paper A Mathematical Theory of Communication published in the *Bell System Technical Journal* (1948). He was trained as an electrical engineer and mathematician. He developed the sampling theorem independent of Nyquist. He is best known for his computation of the available capacity of a communication channel. He obtained his Ph.D. at MIT for his analysis of a mechanical compute engine to compute the solutions of ordinary differential equations.

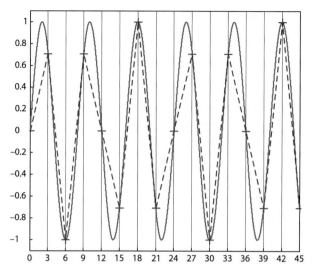

Fig. 4.14. A sinusoidal continuous time signal with period 8 sampled with a sample period 3 yields a periodic discrete time signal of period 24

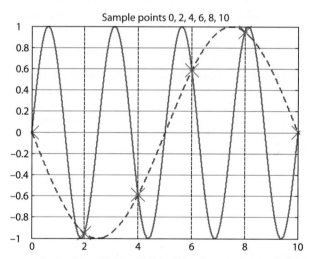

Fig. 4.15. The sine function $\sin(2\pi f_1 t)$ with frequency $f_1 = 0.4$ Hz and the sine function $-\sin(2\pi f_2 t)$ with frequency $f_2 = 0.1$ Hz yield the same sampled data sequence when sampled at 0.5 Hz

Periodic Sampling of Signals: the Nyquist-Shannon Criterion

Periodically sampling a signal without loss of information requires a sampling frequency equal to twice the largest frequency contained in the signal's spectral content. In practice this is enforced through the use of an anti-aliasing filter. This filter ensures that the signal to be sampled at frequency f_s has a spectral content below $0.5 f_s$.

4.5 Signal Processing

Reconsider Fig. 4.12 and denote $y = M(u, d)$. There is always a difference between the signal that is the output of the measurement system y and the signal of interest that was an input to the measurement system u. Signal processing is concerned with understanding signals, and in particular is interested in recovering the signal that should have been measured. In other words, signal processing is interested in the question can we find another system SP that acts on y and recovers u such that, for example, the signal $SP(y) - u$ or equivalently $SP(M(u, d)) - u$ is small in an appropriate manner.

Determining u from information about M, the measurement process, and perhaps some information about the disturbance d, and of course given the measurement y is often a difficult, even an ill-posed problem.

One way of making the problem (much) easier is to impose structure on the input. For example let us assume that u can take on only binary values, for example at any instant of time $u(t) = 1$ or -1. This simplifies matters considerably, and allows us to safeguard against noise and imperfect measurements. That this can be done successfully goes a long way towards explaining why we live in a world where information is stored and communicated in digital format. The same is true for the natural world as well. Indeed, the program for life is encoded in DNA using an alphabet of 4 symbols. That copying, duplicating and executing the program encoded in DNA can be done with very high fidelity despite the randomness of many biological processes is clear from the fact that life still exists today.

Coding and Compressing

By way of trivial illustration, assume that $u(t) \in \{-1, 1\}$ is binary. Let the measurement process simply add noise n. So that $y = u + n$. As long as the noise is bounded, for example $|n(t)| < 1$, it is possible to recover the signal u from y by simply selecting the sign of the output y; indeed $u = \text{sign}(y) = \text{sign}(u + n)$. This is illustrated in Fig. 4.16.

In the engineered world, dealing with binary signals is the subject of *digital signal processing*. When addressing questions about the complexity of binary signals we exploit *information theory*, initiated by Claude Shannon in his famous 1948 paper. Complexity theory addresses questions like: How many different signals are there? The question is pre-conditioned on accepting that signals are only observed via a measurement process with a finite resolution capacity. In such circumstances, it is natural to be interested in two particular problems:

1. How do we safeguard a (binary/digital) signal against corruption by other signals?
2. How can we minimize loss of information if we have to reduce the signal representation because for example we do not have enough capacity to store it or transmit the signal.

The first question is addressed using *coding* methods. Coding methods build redundancy into the signal for storage, retrieval and communication purposes. Redundancy, achieved in a standardized manner, helps in safeguarding against noise. Many errors that may occur in retrieval or transmission can be undone by exploiting the known redundancy in the signal.

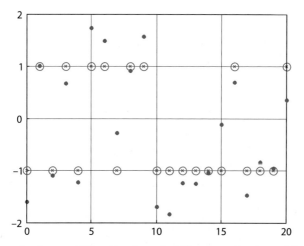

Fig. 4.16. A binary signal recovered from its noisy variant. The measurement error is limited in magnitude by 0.95 in the figure. The original signal is represented by the crosses, and the recovered signal by the open circles and the dots represent the corrupted, measured signal

The second question belongs to the realm of *compression* methods. Here we deliberately reduce the number of bits we use to represent a signal, by doing this in such a way that the essential information is as much as possible retained. Often it is possible to reduce the bit length of a signal by exploiting natural redundancy in the signal. These methods are called *lossless* compression. Lossless compression cannot in general achieve a great reduction in bit length, but it can be undone. Lossy compression methods on the other hand can reduce bit length significantly, but the operation cannot be undone: information is lost.

4.6 Recording and Reproduction

Of course the physical world is not digital to begin with. There are a lot of signals that are not digital, like music. In order to store music digitally we need to build systems that link from the analog world to the digital world and back again.

The main idea is really that we have used a set of basis functions, here the sinusoids with harmonically related frequencies, to represent a measured signal. The economy of this representation and its ability to cope with measurement errors explain its utility as illustrated in the above example. In different signal environments a different set of basis functions can be more appropriate, but the same ideas will apply. There is an interesting trade-off here. By enriching the set of basis functions we obviously are able to model more, but at the same time measurement errors will be able to be modeled as well, and hence we risk to be more sensitive to measurement errors. The fewer basis functions we allow for, the less we are able to model but at the same time we can reject a lot more. From the scientific method point of view, the latter is to be preferred.

One instance where sinusoids are definitely not the right function class to model with is when the signal is a spike train, mostly zero, but once in a while a peak occurs.

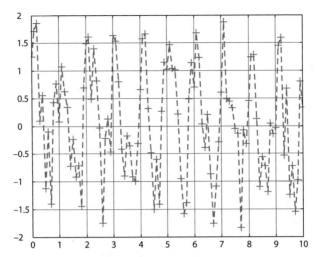

Fig. 4.17. A sine function $\sin(2\pi t)$ sampled with sample period $T_s = 0.1$ with a uniform random measurement error in the range $(-1, 1)$

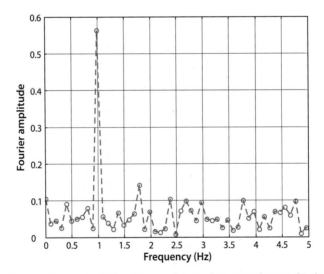

Fig. 4.18. The fourier series representation of the sampled signal Fig. 4.17. The period is clearly identified

(These signals are often used to model neural activity in a brain.) Such signals have a very broad spectrum, and as a consequence the time domain representation is far more compact than is the frequency domain representation. This observation is a manifestation of the *uncertainty principle* associated with the Fourier representation: when a signal's energy is localized in a small subset of the time domain then its spectrum occupies a large subset of its frequency domain (and also when a signal's energy is localized in a small subset of the frequency domain, than its energy is spread over a large subset of the time domain). This is illustrated in Fig. 4.19.

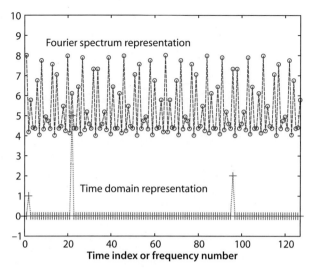

Fig. 4.19. The broad spectrum of a spike train (128 time samples, 128 fourier terms)

4.6.1 Speech Recording and Reproduction

These ideas have a clear interpretation in dealing with playing and recording music or just treating human speech, as discussed in the first chapter. In a sheet of music the composer specifies both time intervals (the beats) and frequencies (the notes) in order to define the music signal. The shorter the note's time interval the less accurate its spectrum can be defined, and the less accurate the note has to be specified. The longer-in-time notes can be much better localized in frequency and therefore need to be produced more accurately to achieve the desired effect. The problem is not that pronounced as for most notes the time interval is actually quite long compared to the period, so the frequency specificity is good. The uncertainty principle is however clearly linked with the way the music scales work, as it determines how many notes can sensibly be distinguished.

By way of illustration of the power of modeling signals with sinusoids, consider speech or music coding. The key observation is that a sinusoid is a rather complicated time function, but it has a point support spectrum. In other words, a complex time function can be stored as a single frequency point, and to be complete as an amplitude and phase. That is an important reduction in the amount of storage required to represent the time function. For example, in creating a digital piano sound, one could record the sounds generated by hitting a key and store these for later play back. Much more economical is to sample the sound, identify the spectrum of the sound, and store the spectrum information, which can be used to regenerate the sound as required. This combined with an understanding of the human ear as a sound transducer helps us to create very economical representations of speech and music.

Figure 4.20 indicates the typical sensitivity of the human ear. The audible spectrum ranges from about 50 Hz to 20 kHz, i.e. the human ear is able to perceive a pure tone of such frequency. This explains for example why CD audio quality sound is sampled

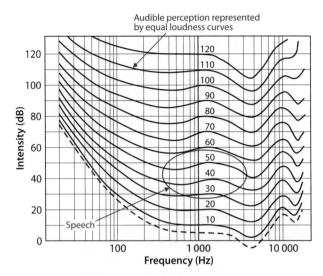

Fig. 4.20. The human ear is an excellent sound transducer with a large frequency and large amplitude range (the amplitude range is presented in decibel $20\log_{10}|P/P_0|$, where P_0 is the reference pressure for audible sound 10^{-5} Pa)

at 44.1 kHz with a resolution of 16 bits (that is $2^{15} = 32\,768$ different levels of sound) over two audio channels. The Nyquist sampling criterion indicates that we need at least to sample at 40 kHz if we want to be able to identify and represent a 20 kHz sinewave. This is a theoretical limit, which can be reasonably well approximated. The practical CD coding is only about 10% above the absolute lower limit, which tells us that the coding technique employed in CD recorded sound is quite powerful.

As indicated in the Fig. 4.20 most of speech is restricted to the frequency band of about 400 Hz to 8 kHz. For telephony purposes it suffices therefore to sample at about 16 kHz.

Of course speech is not a periodic signal, yet analyzing it using the Fourier series ideas underpins much of audio processing. For example in speech processing, speech is first sampled at a high sampling rate, say 25.6 kHz. Then one considers slices of 20 ms in duration. Each of these slices (512 samples in length) is then considered as a period of a periodic signal and equivalently represented using the discrete Fourier series to analyse its spectrum. The spectra obtained in this manner are useful to distinguish and then synthesize the different sounds used in speech. See for example Fig. 4.21, where the sound *bet* is represented in time-domain and every 20 ms time slice is represented in frequency domain. Such a representation is also called a time-frequency representation. The usefulness of the process stems from the fact that the amount of storage space required for the spectra is a lot less than the equivalent time domain sampled signal. The signal's energy is spread out in time domain, and more localized in frequency domain. This process can be used in reverse to synthesize female or male sounding voices that read out an arbitrary input text. A similar process can be used to create a digital piano.

The economy afforded by the frequency representation of the speech signal can be further combined with knowledge of the sensitivity of the human ear. Some frequencies require a much louder sound before the ear perceives the sound (as indicated in

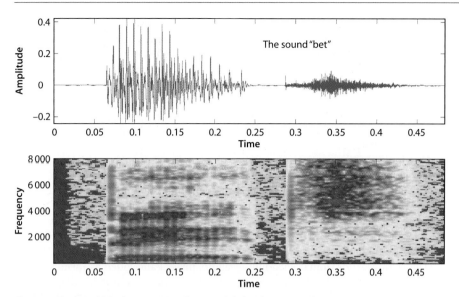

Fig. 4.21. The sound bite *bet*, represented as a sampled pressure wave (on top) and in frequency-time domain (on the bottom). Every 20 ms a new spectrum is identified. The darker areas correspond to higher values in the amplitude

Fig. 4.20). By using less bits of information for those frequencies the ear is less sensitive to, the MP3 coding technique can be extremely economical in storing sound information without sacrificing audio fidelity (it is about 12 times as efficient as normal audio-CD encoding, using the same sampling frequency).

Very similar ideas are behind the coding of image and video signals, like in JPEG and MPEG standards. These standards essentially represent an image using the (spatial) fourier spectrum and truncates the spectrum. The compression that can be achieved in this manner is impressive.

4.7 Comments and Further Reading

Signal processing is an important field of research in its own right. We certainly did not do it justice, but simply wanted to show some of the potential of the frequency domain. The basis functions to understand the frequency domain are the sinusoids and their linear combinations may be used to represent arbitrary signals with any desired level of precision. This is key in dealing with (complex) signals: understanding a few signals really well is sufficient: *divide et impera*.

There are entire books devoted to the topic of Fourier analysis, and the theory has been taken well beyond the realm of dealing with *simple* signals. Not for the faint hearted, Körner's treatment of Fourier Analysis is a comprehensive relatively modern treatment of the topic (Körner 1988). Fourier representation is found in most text dealing with signal processing or signals and systems e.g. Lee and Varaiya (2003).

The library on *digital* signal processing is enormous (Ifeachor and Jervis 2002; Mitra 2005). A good introduction that does not take the Fourier route is Mallat (2001). There the wavelet transform is at the heart of signal processing. Wavelets still use the idea of representing a signal as a combination of a set of basis functions, however the basis functions are carefully selected to minimize the complexity of the representation. Wavelets are introduced in Daubechies (1992). There is an entire industry of wavelet-based signal processing texts.

Because signals are always imprecisely measured, uncertainty, randomness and hence statistics play an important role. Statistical signal processing is also well endowed with texts expounding the key ideas, for example Kay (1993, 1998), Poor (1994) and Stark and Woods (2002).

Viewing signal processing as an inverse problem is surprisingly rare, and this is an aspect of signal processing that certainly deserves further attention.

The information theoretic aspects of signals, including coding are essential in such applications as (tele)communication, or storage and retrieval of signals. One of the key papers is still Shannon (1949). A very readable introduction can be found in Ash (1965). An interesting discussion of coding of signals in the animal and human world is Hailman (2008). A more mathematical treatment is Roth (2006).

The standards for digital encoding of audio, images and video as formulated by the International Standard Organization (ISO, *www.iso.org*) are commercial products and can be explored (bought) via their www-sites. A lot about MPEG may be found at *http://www.chiariglione.org/mpeg/*. MP3 is discussed at *www.mp3-tech.org* and also at *www.mpeg.org*. Most of these techniques now use wavelets specifically tuned for the application at hand.

The ideas around speech and sound encoding, and their realization in the human ear, play a critically important role in signal processing which is at the core of the bionic ear (see e.g. *www.bionicear.org*). The differences and similarities between signal processing as practiced in engineered systems and the neural biology implementation of signal processing is a fruitful avenue for biomimetic research (Bar-Cohen 2006). In fact the MP3 coding technique is one of the most successful biomimetic technologies in the world today.

Chapter 5: Systems and Models

5.1 Introduction and Motivation
5.2 Systems and Their Models
5.3 Interconnecting Systems
5.4 Simplifying Assumptions
5.5 Some Basic Systems
5.6 Linear Systems
5.7 System Analysis
5.8 Synthesis of Linear Systems
5.9 State Space Description of Linear Systems
5.10 A Few Words about Discrete Time Systems
5.11 Nonlinear Models
5.12 Comments and Further Reading

Chapter 5

Systems and Models

> *As far as the laws of mathematics refer to reality,*
> *they are not certain,*
> *and as far as they are certain,*
> *they do not refer to reality.*
>
> Albert Einstein[1]

5.1 Introduction and Motivation

In the previous chapter the emphasis was on signals, how they are generated, analyzed, classified, stored, transmitted, modified, and so on. This chapter looks more carefully at systems and how they act on signals or indeed create signals.

So far we encountered already a great variety of different systems.

- *Signal generators* are systems conceived to produce particular signals. For instance, a chain of integrators generates a polynomial signal. A neural network can represent a large variety of signals depending on the network configuration. A magnetron generates micro waves of a certain frequency and energy content, that can cook food or serve as radar signals. Our mobile phones use signal generators to enable the wireless transmission of our voice.
- *Signal processors* are systems that modify particular features of their input signal. The output signal is determined by the input and the processor parameters. For example a *filter*, or an *equalizer* on our audio equipment adjusts the signal by changing its spectral content, diminishing, or augmenting certain frequency bands as desired.
- *Sensors or transducers* are systems that measure particular signals to represent them in a different physical domain. Their main task is to transform the measured signal generated by a different system into a more suitable signal often belonging to a different physical domain, so that it may be stored or processed further. The radio telescope from Chap. 3 is a sensor system, and so is our skin.
- *Receivers and transmitters* are systems that form part of any communication system. The transmitter converts a signal into an appropriate analog domain suitable for transmission, the receiver takes the analog signal and attempts to recover the signal that was transmitted. Our voice (the transmitter) and ears (the receiver) form part of our personal communication system, where our brain plays the role of signal processor.

In all of these examples, the systems are subservient to the signals. Nevertheless all together, systems are not there just for the signals, e.g. they may serve the more important purpose of transforming materials and energy.

---

[1] Albert Einstein, 1879–1955, German-born theoretical physicist, one of the great minds of the world. He is best known for the theory of relativity. He received the 1921 Nobel Prize for physics for the discovery of the photoelectric effect.

When dealing with signals, we brought order by considering signals as constructed from a collection of more basic signals. The same holds for systems.

First observe that from our discussions in Chap. 2 we already know that we can abstract away from the physical domain of interest for the purpose of modeling, and design. As it turns out there are a number of elementary building blocks with which we can construct arbitrary system models. Understanding these elementary systems provides us with a lot of insight. Then we need to learn how we can build larger systems. This is achieved through the interconnection of the (elementary) systems, either in cascade, parallel or in feedback. How this must be done in order that the system achieves its purpose, is the subject of design. Interconnecting systems allows us to build systems of much greater and increasing complexity rather quickly. As a consequence, characterizing the resultant behavior from our understanding of the elementary building blocks quickly becomes intractable. The development of tools that allow us to understand the emergent behavior of a large system and that allow us to explore the various scales of large systems is the subject of much research in the systems engineering community. Interestingly, and more often than not, the fine detail of the internal workings of all the subsystems is not necessary to understand the overall behavior of a large scale system. Feedback plays an important role in this emergent simplification.

This chapter is organized as follows: first the concept of models is revisited. Particular attention is devoted to the idea of a system being considered as a (mathematical) operator and the facilities this will provide to deal with complex systems. Interconnection of systems is considered next. Special classes of models are touched upon. In the last section, some thoughts about modeling are presented.

5.2 Systems and Their Models

From an abstract point of view, models are systems representing the system of interest in a different more convenient manner. Architects use scaled models of a new building, both to help their clients to appreciate the design, but also to demonstrate the aesthetics, the functionality and understand how the building works. A model could be a computer rendered drawing as well. When dealing with systems where signals change dynamically over time, the most convenient models are computer algorithms that reflect the system's behavior.

Practically[2] there are two main ways of getting to a model, from observations of the signals derived from the system, or from breaking the system down into a collection of interconnected subsystems for which we already have a model. Often, both methods are combined to arrive at a model.

5.2.1 From Signal to System Model

It is through the observation of signals that the nature of a system is revealed, see Fig. 5.1a. At this level of abstraction a model is equivalent to its *behavior* or the collection of *all* the signals that are compatible with the system under consideration.

---

[2] Philosophically, there is only one way in which we obtain models, that is from observations through signals. If we believe to already have a model for a system, we are merely relying on someone else's observations.

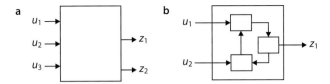

Fig. 5.1. A system: **a** observed through inputs and outputs; **b** its structure

In some simple systems (like an automaton), the behavior is easily obtained and described. For instance, consider a room with a single light commanded by a single light switch. There are two possible values the input (position of the switch) can take, on or off. The output is our observation of the light, also either on or off. Even in this simple case we must make the assumption that there is power and that the light bulb and/or switch is not faulty. The behavior of this system would then consist of all sequences where both input and output are equal. If we had an observation reported "switch is on and light is off", we could decide that either there is a fault, or that the observer made a mistake.

In general, to observe the collection of all signals is all but impossible to obtain, but we may be able to conceive to be in possession of all possible observations from a system, by extrapolating from our past experience.

Notice that we allow explicitly for the situation that there are other input signals into the system as well, other than the ones we observe. They are considered part of the internal workings of the system. Of course not knowing all the inputs may pose some problems when we are interested in explaining what is actually happening. Think about the light in the room, there is not only the room switch but also the power supply.

When approaching a system from the point of view of the signals that can be extracted from it, we explore systems very much in top down fashion. We start with the global system, and continue exploring what is inside through what can be observed from the signals. What do these signals actually reveal about the system? Can we unravel some of its internal structure? Can we detect feedback loops? How can we use the available inputs to probe the workings of the system? For example, can the inputs that we are able to exploit dominate the input signals we are unable to manipulate or observe? These questions go to the heart of *systems identification*, which is the study of how to move from signals to mathematical equations that describe the system.

In order to arrive at a useful mathematical description we need to take the step from signals to equations. This is often called *from data to model*. Although the latter is a bit of a misnomer, as our data collection is itself the best model we have for the system. It is perhaps an unwieldy model, but a model nevertheless. The main aim in modeling is of course to arrive at a more compact description, preferably a set of equations or even better a computer algorithm that summarizes our observations and could be used instead of the actual system for further experimentation and exploration.

In this search for a set of equations, we should keep in mind why we need a model (there is no such thing as *the* model). The purpose of a model is important. A process model can be used to synthesize other components (for instance, a controller), to better understand a process (to teach someone how to work with the system), and to simulate it or as a tool to improve its design. Also, a model of a process is always a partial representation of its behavior, only valid for a given range of signals.

Are we interested in simulation to see the system response for particular inputs? Are we interested in designing a feedback loop, to enhance certain system properties and eliminate unwanted behavior? It is clear that in the former question we will need details in the equations that allow us to explore all the inputs of interest, but no more than that either. In the situation for feedback design, we only need a model to the extent that it allows us to design the feedback well. The unwanted behavior, the one that we suppress through feedback, is in this instance not required in great detail for this model. Of course a model that can be used with all possible inputs will do, but in a feedback situation, we cannot see all possible inputs as some inputs are going to be subject to feedback (the next section will make this clearer). The purpose a model is used for significantly determines the type of model we need.

Typically we take this step from signals to equations, by postulating a class of equations, or models, indexed by some free parameter(s). Then we try to select from all these equations those equations that fit the signals we have observed in some best possible way. The idea goes back to Gauss[3] who tried to fit the astronomical observations of the planets in our solar system to Kepler's model of planet motion: orbits are ellipses with the Sun in one of the foci. In this method the data are put into the models, the misfit is computed for each model (each parameter selection), and the model that produces the smallest error is declared *the* model. Preferably the best error should be smaller than what is indicated by the typical measurement error in the signals, otherwise we should select a richer model class. The same principle is still used today, and there exists substantial literature dealing with all the possible variations on this basic theme. Substantial computer software packages for computer assisted modeling from data has been developed.

5.2.2 From System to Model

An alternative approach to finding a model is to view a system as a network of interacting subsystems, see Fig. 5.1b. An immediate example is an electric circuit composed of resistors, capacitors, transformers, transistors and so on. We know the behavior of each component as well as the constraints imposed by the network connections. Using these building blocks a model of the circuit can be derived in a systematic way. In this bottom-up approach we infer the system behavior from our understanding of the elementary subsystems and their interconnections.

This line of modeling is very useful in analysis and also in design of engineering systems as in this case we (should) know what the various building blocks are as presumably each subsystem has been thoroughly designed and tested. Moreover, the rules of interconnection are well-defined and ensure that the global system will behave as designed. Most large scale engineering systems are constructed using this bottom-up approach.

Computer software packages dealing with this approach have been developed. Typically these tools are limited to a particular domain of application. For example there is extensive software to deal with models of fluid flow, or for designing a large scale

---

[3] Johan Carl Friedrich Gauss, described in 1806 a method, called least squares to study curve fitting in general. He was an enormously prolific mathematician and ranks amongst the most influential in history. He lived and worked in Germany, 1777–1855.

integrated circuit, or for constructing a bridge or ship. Packages that bring several domains together are starting to emerge. An even more recent development is to also bring biological subsystems into the fold of this modeling paradigm.

The main issues associated with this approach are scale and reliability. As the number of components or subsystems grows, so does the potential for failure in subsystems. As the probability of some failure somewhere in the system grows with the number of components the incidence of failure becomes a certainty, and the model for each subsystem must capture normal as well as abnormal operational conditions. This implies that there is no longer a single model for the overall system, but a (large) collection of alternative models. The complexity of such models grows much faster than the number of components, and often becomes intractable from a purely computational point of view. Measurements to diagnose which condition applies, so that it is possible to determine which model is the right one and feedback to take appropriate action, for example changing the interconnection pattern, play an important role in making such large scale systems work in practice.

Totally new designs require combinations of top-down and bottom-up approaches. Understanding simple systems is essential in building our intuition on how larger systems do work, or may work. A top-down approach is necessary to ensure that we stay focused on building a system that delivers what is expected. Also in exploring the natural world we typically combine both top-down and bottom-up approaches.

5.2.3 Model Classes

Model classes are often distinguished on the basis of which signals take center stage. We distinguish

- *Input/output models.* Sometimes also described as external models, they represent the process as a *black box* which merely expresses dynamic relationship between the input and the output signals. They rarely can do so only using the input and output signals, and more often than not models will introduce additional signals to obtain the input/output descriptors.
- *State space models.* An internal state of the model is the key variable. The system input acts as to influence the state in the model, and the output is derived from the state. State space models are very common, and most simulation packages work in terms of state space models. (Of course they define an input/output relationship as well.)

We can focus our attention on some signal characteristics, or the way we represent time. If time is not important at all, but rather the relative order in which *events* occur in the system, we speak of *event-based models*. Signals can be stochastic, or set-based, or deterministic, discrete and continuous. The list of possibilities is rather endless.

If the signals are not represented in time domain, but rather in frequency domain, we speak of frequency domain models. Sheet music is an interesting modeling language, using a mixture of beats and frequency to represent music; precise, and yet with sufficient latitude to allow for various interpretations.

If all signals are binary the model is logic-based, and the modeling language uses the rules of Boolean algebra. Automata, like the car washing system described in Sect. 3.5 form a class of such models.

> **Boolean Algebra and Logic Models**
>
> A binary (Boolean[a]) variable can only take one of two possible values, say either 0 or 1. The result of any boolean operation on binary variables is again a binary variable.
>   The basic boolean operators are:
>
> - AND: (A AND B) the result is 1 if both the variables A, B are 1
> - OR: (A OR B) the result is 1 if at least one variable A OR B is 1
> - NOT: (NOT A) the result is the opposite to the argument variable A
>
> A logic model will be composed of a set of equations such as:
>
> $L = M1$ AND $X3$
> $\ldots = \ldots$
> $F = L$ AND $X1$ OR $X3$ AND NOT($M3$)
>
> [a] George Boole, 1815–1864, English mathematician and philosopher, best known as the inventor of Boolean logic, and hence can be considered in hindsight, the founder of the field of computer science.

We can talk about local or global models, depending of the range of variables we want to consider, or the modes of operation we want to include in the model. Local models may not include all the failure modes for example, whereas a global model may. So for example, an aircraft's auto-pilot may be designed on a local model, but the pilot is trained on a global model for the aircraft's behavior.

Most models only represent a portion of the entire behavior. The latter is normally too complex, and in most control synthesis questions, only a small portion of the entire behavior is required or considered.

Models are therefore often simplified, common simplifications are:

- *Linearization*: where we restrict the range of the signals, only considering small deviations from a nominal signal. If the signals are indeed restricted to be small, it is possible to enforce a linear model. This simplifies greatly analysis and synthesis. In a control context, where the main task is often to regulate signals near a desired reference signal, linearization is often the first simplification that can be used.
- *Reducing complexity*: Models may become very large if all possible phenomena at all time/frequency scales are taken into account. By restricting oneself to either very short time scales, or very long time scales, or excluding a part of the behavior, a model can often be considerably simplified. For example, in systems dealing with speech we may restrict models to those frequencies in the human speech spectrum, ignoring any other frequencies that can be perceived. By way of another example, in trying to model the human ear, we concentrate only on the rest state, avoiding the complexity of cognitive processes.
- *Reducing dimension*: (Another form of complexity reduction) by simply ignoring part of the state of the system, and perhaps dealing with the ignored part as a disturbance signal from the environment. So some of the dynamics become simply an external input signal. For example in dealing with the dynamics of tides, we may want to include the Moon, and perhaps the Sun, but are quite willing to ignore all other planets. More drastically, we may simply cut off a whole range of possibilities

because we want a simple model. For example in dealing with the temperature in an oven, we could model the temperature as being spatially distributed across the oven, (as it really is) but if we only have one input, say fuel flow, there is no way we can control the temperature distribution, but we may be able to control the mean temperature in the oven. So the oven is simplified to be represented by a single heat storage device, characterized by a single temperature.

> **Models**
>
> A model is a partial representation of a system's (dynamic) behavior. There is no *the* model for a system. Many different models can be associated with the same system depending on what level of approximation we desire. The latter is a function of the purpose for the desired model. A model should be represented with a quality tag indicating its fidelity in reproducing the system's behavior, or the range of signals it is valid for, or the size of approximation error we may expect.

5.3 Interconnecting Systems

The basic interconnection that we allow for in our discussions is that an output of one system becomes an input into another system. The idea is abstractly described in diagram Fig. 5.2. An interconnection assumes that the signals that are *made common* are at least compatible in that they must have the same range and domain.

In the new system that results from interconnecting two systems, a major change is that the number of free variables has decreased. Indeed in order to have an interconnection at least one input must be replaced by an output of another system. In Fig. 5.2, one output, z_1 has been prescribed as the input, v_1, of the other system. The new interconnected system has fewer inputs than the number of free inputs that were available before the systems were interconnected. The number of outputs does not have to change. Indeed, whether or not the signal z_1 is considered an output for the newly interconnected system in Fig. 5.2 is a mere matter of choice. Given it

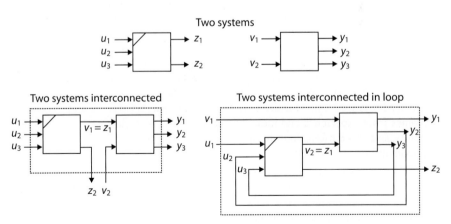

Fig. 5.2. The basic system build construct: interconnecting systems implies a loss of degrees of freedom

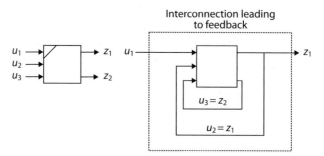

Fig. 5.3. Creating feedback loops implies a loss of freedom. At least one input is replaced by an output

was available for measurement before the interconnection was made, we may as well assume it remains available for measurement after the interconnection, but this is not essential.

We may also consider interconnecting outputs with inputs on the same system. In so doing we create *unity*[4] feedback loops. Again, the main feature of the interconnection is that we have reduced the number of free signals. The idea of creating feedback loops is illustrated in Fig. 5.3.

In block diagrams like Fig. 5.2 or 5.3 the arrows play a significant role in that they identify the direction of information flow, or causality. Systems act on inputs to produce outputs. When confronted with the task of modeling a particular system it may not always be feasible to distinguish inputs and outputs clearly (we come back to this in the next section). This can be due to the fact that we are simply taking measurements, not knowing the causality relationship, a situation not uncommon in econometrics or a consequence of the complexity of the system under consideration. We can develop the idea of an interconnection without reference to causality in that an interconnection is simply equating signals. That is, two signals related to two subsystems are forced to be the same by means of an interconnection, but this will not be pursued here.

The precise behavior prediction of a complex system knowing how the individual building blocks behave turns out to be surprisingly difficult, although in the special case of linear systems the situation is quite well understood and there are excellent tools available to explore their behavior. Not only analysis but also synthesis issues are well understood in the linear case. Linear systems are rather special and most systems exhibit nonlinear characteristics. In this more general context, our ability to guarantee certain behavior, or perhaps better to exclude with certainty some undesirable behavior is very limited indeed. In engineering systems over-all system reliability comes from experience and judicious use of feedback to eliminate or reduce uncertainty and more often than not a lot of insight is gained by considering linear approximate models.

---

[4] Unity in the sense that there is no other system intervening in the loop, or the unity operator takes a signal and simply reproduces it at its output.

5.4 Simplifying Assumptions

Linear, causal and time-invariant systems (see Sect. 1.5) form the class of systems that is most studied in literature.

In modeling a process we start from the collection of all signals, inputs and outputs that are compatible with it. This collection is called the *manifest behavior* of the system. Although perhaps too general a framework, the concept of a behavior allows us to define rather precisely what we mean by causality, time invariance and linearity. These properties simplify the analysis and indeed synthesis of systems considerably. Unfortunately, most systems are strictly speaking neither time invariant nor linear. Nevertheless, by appropriately restricting the operation of a system, for example through the use of feedback, it is often possible to ensure that both time invariance and linearity hold with sufficient degree of accuracy enabling their use in analysis and design.

Tools for analysis, design and synthesis of linear, time invariant systems are well-developed, and include computer aided design environments that can efficiently deal with truly large scale systems.

In the nonlinear system setting things are much less-developed, and more often than not linear tools are used in an iterative manner to unravel the complexity associated with nonlinear behavior.

Causality

Our experience indicates that we live in a causal world, actions cause re-actions. Reactions cannot anticipate actions. Nevertheless establishing causality as an inevitable part of the world as we know it is not a trivial matter (Zeh 1992). For systems, as we discuss them, where time is the essence, causality is a natural property to postulate.

It is easy enough to mathematically formulate systems that are not causal. Consider the following system, with input u and output y, and where y is a moving average of the input as follows:

$$y(t) = \frac{u(t-1) + u(t) + u(t+1)}{3}$$

This system is clearly not causal. The present output $y(t)$ depends on $u(t+1)$ the input at a future time. Such systems are excluded from our consideration.

> **Causal Systems**
>
> We say that a system is causal provided that the current value(s) of its output(s) do not depend on future values of its input(s). We say that a system is strictly causal if the current value of the output does not depend on either the present nor the future values of the input(s).
>
> If we did not know a priori what signal is an input and what is an output, we could use the above test to separate signals into inputs and outputs and establish a causality structure.

Time Invariance

Loosely speaking a system is time invariant if its behavior does not explicitly depends on time; or experiments performed with the system will yield the same answers regardless when they were performed. With a little more precision, we could say something like

A system is time-invariant provided that for any valid input-output pair (u, y), that is an input-output pair belonging to the manifest behavior, any time-shifted version such as $(u(t+\tau), y(t+\tau))$ (for some scalar τ) is also a valid input-output pair, i.e. belongs to the manifest behavior.

Clearly time-invariance is more easily decided in the negative than in the affirmative, as the former requires but one counter example. We will normally assume that the systems we observe are time invariant. It is after all a very cheap assumption.

> **Time-Invariant Systems**
>
> We say a dynamic system is time invariant if regardless of the time instant an input is applied, the same response is obtained.

Linearity
A system is said to be linear if for all input-output pairs (u_1, y_1) and (u_2, y_2) that belong to the (manifest) behavior of the system, the linear combination $\alpha(u_1, y_1) + \beta(u_2, y_2) = (\alpha u_1 + \beta u_2, \alpha y_1(t) + \beta y_2)$ is also a valid input-output pair (that belongs to the manifest behavior) for all scalars α and β.

As observed, linearity is rather rare. Nevertheless, most practical systems can be considered linear in some small region of their normal operating regime (i.e. the property of linearity holds not for all input-output pairs, but for a reasonably sub-collection of input-output pairs). Also, understanding linearity is a necessary step towards understanding nonlinear systems.

> **Linear Systems**
>
> A system (an operator) is linear if its response (outcome) fulfils the principles of superposition, addition and proportionality with respect to all the possible inputs.

One interesting feature is that these three system properties remain unaltered (they are invariant) when we interconnect systems that have these properties.

That an interconnection of linear systems is again linear is perhaps obvious when considering that an interconnection simply imposes a constraint in the form of an equality of signals. By its very nature an equality respects scaling and additions, hence linearity.

An interconnection of (strictly) causal systems is again causal. The equality constraints imposed by the interconnection does not alter the time dependent relationship between remaining inputs and outputs. Perhaps the feedback interconnection requires a little more reflection. Here an output that causally depends on a particular input becomes causally dependent on its own past through the interconnection. That is why we imposed to work with strictly causal systems. Otherwise algebraic loops can be created, and it is not clear if these can be resolved or not. If no feedback loops are created in the interconnections the requirement for strict causal subsystems is not necessary.

Clearly, an interconnection of time-invariant systems is again time invariant.

5.5 Some Basic Systems

As indicated, a system can be viewed as composed of interconnected (elementary) systems. Some of the simplest systems encountered in feedback and control are described next. The list is by no means exhaustive.

5.5.1 A Linear Gain

The simplest system or model is perhaps a linear *gain*: the output signal y is simply the input signal u multiplied by a scalar K, often called the gain: $y = Ku$.

The response when the input is a sinusoid is thus a sinusoid of the same frequency and the same phase but of different amplitude. The output amplitude is the input amplitude times the gain. The typical block diagram used to represent such a system is given in Fig. 5.5a.

Some examples of physical systems that are well-approximated or modeled by a pure gain are mechanical systems such as a lever or gearbox or an electrical system like an audio amplifier.

Consider the lever in Fig. 5.4. The signals we measure on the lever are the displacements at either end. Here denoted as u and y. What is input or output depends very much on how the lever is used. At the level of describing the behavior this is actually irrelevant, we do not have to decide what signal is input or output (in this case we are in fact not able to make the distinction anyway). In all situations we will find that the displacement signals are proportional to each other. The lever is a linear system, time invariant, and causal but not strictly causal. With the notation reflecting the situation in Fig. 5.4, the behavior is the collection of all signals u, y such that

$$\frac{y}{\ell_y} + \frac{u}{\ell_u} = 0$$

where ℓ_u and ℓ_y are the lengths of the lever from the fulcrum to the edges where u, y are measured respectively.

When one side of the lever is actuated say the u-side, and the other side is used to move a load, the y-side, it becomes clear which side of the lever should be considered the input, or cause, and which is the output or effect.

Other measurements or signals that can be associated with the lever are forces.

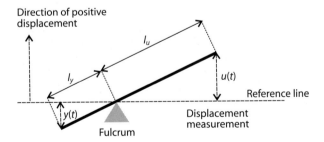

Fig. 5.4. A lever, with displacement measurements u and y

Combining forces with displacements we could consider building a dynamical model for the motion of the lever using its inertia and so on.

A pure gain system does not have memory. In this case, the output equals the input up to a scalar, so all we need for predicting the future of the output is the future of the input. There is no state variable in this case.

A similar example is a gearbox the input could be the speed of the input shaft (motor driven shaft), the output the speed of the output shaft (the load side). An alternative input may be a force or torque on the input shaft and the output could the output shaft position, velocity, or perhaps the torque on the output shaft. The same physical system may have many input-output pairs associated with it. The choice reflects our interest, or better the transducers that are in place and really these are an integral part of the physical system under consideration.

In an audio amplifier the input would be the electrical signal from say a compact disk, or microphone, the sound produced by the speaker is the output. Such a system is however not a passive system. The total (electrical) power available at the input to the audio amplifier is typically much less than the (sound) power available at the output. The fact that we want to represent the system by a linear gain in this case indicates that we are not really interested in the inner working of this particular system. This is often the case for transducers, sensors or actuators. The latter normally provide large power or energy gain as well as signal gain.

When we model a transducer as a linear gain, the gain must account for the adjustment of the physical units from input to output. For example in the audio-amplifier, the input signal will typically be an electric voltage (measured in Volts), and the sound signal is expressed in units for pressure (measured in Pascals).

In practice there is no system that is simply a linear gain. There will always be some physical limits within which the system has to operate. A lever will bend when too much force is applied. The speakers will saturate and distort the sound when too much power is requested, or when too high a frequency signal has to be represented. Other effects may occur for very small signals. In the example of the gear train, the output response will lag the input when reversing the direction of rotation (e.g. see Sect. 3.4). Or there may be no response at all if the input torque cannot overcome the friction and stiction in the gear train. When these effects are important they should of course be part of the model. When we use a linear gain model for a particular system, we have made the implicit assumption that the signals presented to the system/model are such that these complicating effects may be ignored.

The linear gain system is characterized by two important properties: instantaneous (no memory) and linearity.

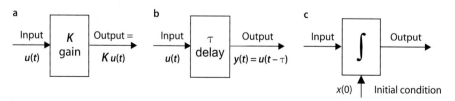

Fig. 5.5. Basic systems: **a** static gain, **b** pure delay, **c** integrator (accumulator)

5.5.2 Transport Delay

Another frequently encountered phenomenon is that of *delay* (Fig. 5.5b). The sound we hear is a pressure wave that had to travel from the source to our ears. This takes time (distance divided by the speed of sound).

Look at the transportation belt in Fig. 5.6. The raw materials, clay and lime, flow out from the hoppers and lay on the belt. They are transported to the mill. Obviously, what enters the mill at a given time instant is what entered on the belt some time ago (see Sect. 3.2).

A simple system representation for this is

$$y(t) = u(t - \tau) \tag{5.1}$$

The output signal equals the input signal time-shifted by a delay of $\tau > 0$ units of time. A delay system is dynamic, strictly causal and linear. We will find it useful to rewrite the above equation as follows:

$$y = \Delta_\tau u \tag{5.2}$$

which we read as the output y is equal to u delayed by τ. Δ_τ is the delay *operator*. When the input is a pure sinusoid the output of a delay system has the same pulsation as the input, and a phase shift with respect to the input that is proportional to the pulsation and the time delay. A delay does not affect the amplitude of the signal. This is exemplified in the next equation:

$$u(t) = U\sin(\omega t) \quad \rightarrow \quad y(t) = U\sin(\omega(t - \tau)) = U\sin(\omega t - \omega \tau)$$

An interesting property of the delay is that it can change the sign of a signal. In the previous equation, if the period of the input is twice the delay $2\tau = 2\pi/\omega$, the output will be in anti-phase with the input, that is $y(t) = -u(t)$.

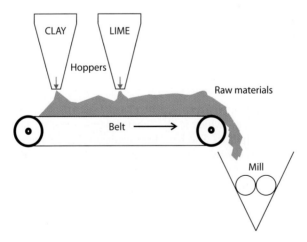

Fig. 5.6. Raw materials transportation belt. Transportation time delay

5.5.3 An Integrator

Another model that is easy to understand and is frequently used to model physical systems encountered in practice is the *integrator*. As seen in Chap. 2, the integrator captures the idea of conservation, or storage. A simple physical example of such system is a reservoir (remember the toilet's cistern, see Fig. 1.2 for an example). The reservoir stores or integrates the difference between inflow and outflow. The level of the reservoir is the important variable here. Suppose we start filling/emptying the reservoir from say time 0, the level at a time t will then depend on both what the inflow/outflow was over the time interval $(0, t)$, and what level the reservoir had at the beginning. We say that the reservoir has *memory*. This memory is also called *the state* of the tank. In practice the reservoir will have finite capacity, it cannot be less than empty or more than full. An integrator is an abstraction of such a reservoir, one without limits (i.e. it cannot be emptied nor filled). Physical examples of reservoirs can be taken from electrical engineering where capacitors can be modeled as integrators (capacitors are reservoirs for electric charge), or hydrology where a water reservoir or bath tub can be modeled as such, or thermodynamics, where an insulated oven is a heat reservoir. In those examples, the various inputs would be respectively the electrical supply, the position of inlet and outlet valves for the fluid and an heat exchanger with a cold or hot fluid able to extract or supply heat. The output would be the level of the reservoir, a measure respectively for the charge stored on the capacitor, the fluid volume in the bath tub or the heat stored in the oven. A schematic representation is given in Fig. 5.5c. We will find it useful to use an operator notation for the integrator as follows:

$$y = \int u \qquad (5.3)$$

More mathematically complete, using some calculus, an integrator with output y and input u and initial condition y_0 is written as

$$y(t) = y_0 + \int_0^t u(s)ds$$

or also as[5]

$$\dot{y}(t) = u(t), \quad y(0) = y_0 \qquad (5.4)$$

We read this last equality as, the derivative of the output equals the input, or the output integrates the input (which is also what the first equality states). In order to specify uniquely the output we also have to provide a value for the output at a given point in time (this is the so-called initial condition y_0). This representation (Eq. 5.4) using the derivative is very useful, and is typically the way it is used in simulation packages for computational purposes.

---

[5] We use the notation $\dot{y}(t)$ to denote the derivative of the signal $y(t)$ with respect to time, other notations are $\frac{d}{dt} y(t)$ or simply Dy.

An integrator is clearly causal, the present output or state only depends on the past of the input.

An integrator is time invariant. Thinking of a water reservoir, supplying water today, tomorrow or yesterday always yields the same change in water volume for the same supply. The equations reflect this as there is nothing that refers to a time dependency, other than the signals themselves.

An integrator is also linear. Indeed if (u_1, y_1) and (u_2, y_2) are in the behavior of the integrator, then so is $\alpha(u_1, y_1) + \beta(u_2, y_2)$ for any scalars α, β. This can be easily seen on either equation (remember, the initial condition!).

It is instructive to see how an integrator acts on a sinusoidal input, $u(t) = A\sin(\omega t)$, starting from an initial condition $y(0) = y_0$.

$$y(t) = y_0 + \int_0^t u(\tau)d\tau; \quad \rightarrow \quad y(t) = y_0 + \frac{A\sin(\omega t)}{\omega} \qquad (5.5)$$

An integrator system has the following properties:

- An integrator has memory, as captured by the initial condition. In order to know the future output, the present output and the input over the future time interval of interest are required. The present value of the output therefore serves as a *state* variable for the integrator system.
- A zero input response of an integrator is a constant signal. So an integrator can be viewed as a signal generator for the constant signal.
- A constant input response of an integrator is a ramp signal. So two integrators connected in series, with a zero input, can be viewed as a signal generator for the ramp signal. (See also Sect. 4.2.1.)
- The integrator response consists of an initial condition response $x(0)$ combined additively with a response due to the input $\int_0^t u(\tau)d\tau$.
- The integrator response at time t depends *causally* on the input. The input has only to be known up to the present (the present is not required) in order to determine the present output.
- An integrator is a linear, time invariant and strictly causal system.
- Let the initial condition be zero, or focus only on that part of the response that is due to the input. For a sinusoidal input, see Eq. 5.5, we have:
  - The output has the same pulsation as the input. This is a property all linear systems possess.
  - The output has amplitude $A \rightarrow A/\omega$. The faster the input changes, the larger ω, the smaller the response. We say that the integrator has a *low pass* characteristic. If the frequency of the input increases by a factor of 10 (one decade up) its gain is reduced to $1/10$, which is usually expressed as a reduction of 20 dB[6].
  - The phase of the output is, $0 \rightarrow -(\pi/2)$, we say that *the output lags the input* by 90 degrees or $\pi/2$ radians.

---

[6] To express a gain K in decibels, we use the expression $20\log_{10}K$. Decibels is abbreviated as dB.

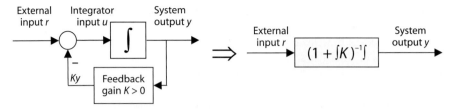

Fig. 5.7. Gain feedback around an integrator. Equivalent input-output representation

5.5.4 An Integrator in a Feedback Loop

An integrator in a negative feedback loop is represented in Fig. 5.7. In Chap. 2, see also Sect. 1.2, we encountered examples of systems that can be modeled as an integrator in a negative feedback loop. All such systems are well-behaved.

It is instructive to see how the system output y is related to the external input r. The block diagram informs us of the following relationships. We start with the summing junction in the block diagram. The input to the integrator u is the difference between the external signal r and the output of the gain block Ky, namely $u = r - Ky$. The output of the integrator is $y = \int u$, and hence replacing u in this expression using the previous expression for u leads to $y = \int(r - Ky)$, because of linearity this is equal to $y = \int r - \int Ky$ or $(1 + \int K)y = \int r$. This may be rewritten as $y = (1 + \int K)^{-1} \int r$. We may think of $(1 + \int K)^{-1}$ as the inverse of the operator[7] $(1 + \int K)$.

The block diagram can therefore be replaced by a single linear system with operator $(1 + \int K)^{-1} \int$, which as long as $K > 0$ is indeed well behaved (see Fig. 5.14). These simple manipulations allow us to eliminate feedback loops from a block diagram, and hence compute from knowledge of the operators for each subsystem readily an operator representation for the entire system.

5.6 Linear Systems

Similar to the integrator in a loop, gains, delays and integrators can be interconnected to represent a rich collection of dynamics with the properties that the behavior will be linear, causal and time-invariant. It is perhaps surprising that most systems can be really well approximated using such models. Even when systems show clear departure from linear behavior, it is possible to obtain excellent linear approximate models for them by considering inputs or signals of sufficiently small magnitude.

In the end, a linear system model is represented by a set of n differential equations (or difference equations in the case of discrete time systems) with linear structure and with constant coefficients, such as:

$$\dot{x}_i = a_{i1}x_1 + \cdots + a_{in}x_n + b_{i1}u_1 + \cdots + b_{im}u_m \tag{5.6}$$

---

[7] This is just a very suggestive notation. Suppose the operator P maps input u to output y. If the operator P has an inverse, the inverse maps from the output back to the input, that is $u = P^{-1}y$, the inverse of P is then denoted as P^{-1}. As always, inversions must be treated with care.

where u_j; $j = 1, 2, \ldots, m$ are the system inputs. This can be expressed in matrix form

$$\dot{x} = Ax + Bu; \quad y = Cx + Du \tag{5.7}$$

where y represents the set of output variables. The vector x collects all the components x_i, x is a state variable. The vector u collects all the input components u_j. The coefficient matrices A, B, C and D collate in a two-dimensional table format all the coefficients a_{ij} and b_{jk}.

Based on the system structure and the operating input/output models of the components, a global operating model of the system, as a black-box, can be easily derived. For that purpose, the algebra related with the block diagram representation introduced in Sect. 1.3 can be used.

5.6.1 Linear Operators

Black-box models, often referred as input/output models, express the processing of the signals in the system by means of an operator. As we are dealing with the special class of linear, time invariant systems, these operators are also linear and can be easily combined using block diagram manipulations.

By using the notation as used with the integrator in the previous section, two main operators can be considered for continuous time systems:

- **Transfer function.** Given a system with input $u(t)$ (with Laplace transform $u(s)$) and output $y(t)$ (with Laplace transform $y(s)$) (see Sect. 4.3), and assuming null initial conditions, the Laplace transform of the output is proportional to the Laplace transform of the input. The factor between input and output in Laplace domain is the system operator:

$$y(s) = G(s)u(s) \tag{5.8}$$

It is typically called the *system transfer function*. In a block diagram, we will often write $G(s)$ inside the box representing the system[8]. It suffices to simply multiply the input with the transfer function to find the output (all in Laplace domain of course).

This illustrates the power of the Laplace domain, what appears to be calculus (taking derivatives) in the time domain is mere algebra in the Laplace domain.

By way of example, consider an integrator. If we use the Laplace transform properties (see also Sect. 4.3), and apply to Eq. 5.4, the result is (always assuming zero initial condition):

$$sy(s) = u(s); \quad \text{or} \quad y(s) = \frac{1}{s}u(s); \quad \text{or} \quad \int = \frac{1}{s} \tag{5.9}$$

---

[8] For a linear system with the representation 5.7, the corresponding transfer operator is given by $G(s) = C(sI - A)^{-1}B + D$. Where $^{-1}$ refers to the matrix inverse.

The inverse operator, the differentiation, is represented by s. An integrator is represented as s^{-1}. Differentiation and integration are each other's inverse. Using s to represent differentiation and the rules about block diagram manipulations goes a long way towards explaining why all linear systems can be represented by some factor $G(s)$ linking input with output.

- **Frequency response.** Here we are interested in how a linear system behaves when the input is sinusoidal: $u(t) = U_0 \sin(\omega t)$. In steady-state the output[9] is again sinusoidal, with the same pulsation, a different amplitude and a phase shift: $y(t) = Y_0 \sin(\omega t + \phi)$. The input-output relationship can be expressed by a complex quantity $G(j\omega)$, such that its modulus and argument are (see box on p. 92):

$$|G(j\omega)| = \frac{Y_0}{U_0}; \quad \angle G(j\omega) = \phi \quad (5.10)$$

Thus, for each pulsation ω_1, the operator $G(j\omega_1)$ provides the gain and phase shift of the output with respect to the input. This can be easily obtained if the exponential representation of the sinusoidal signal is used instead. With an input given as[10] $u(t) = Ue^{j\omega t}$, the system output (in steady state) is given by

$$y(t) = UG(j\omega)e^{j\omega t} = U|G(j\omega)|e^{j\omega t + j\angle G(j\omega)}$$

The notation, using G as for the Laplace transform, is suggestive of a particular relationship between the transfer function and the frequency response characteristic, which is indeed:

$$G(j\omega) = G(s)\big|_{s=j\omega} \quad (5.11)$$

Transfer Function Operator

The transfer function operator for a linear time-invariant system provides an algebraic representation for linear systems in the frequency domain. The system's response y in the frequency domain is simply the product of the transfer function $G(s)$ with the input u (in frequency domain) that is $y = G(s)u$. For a sinusoidal steady state, with input $u(t) = Ue^{j\omega t}$ the response is $y(t) = G(j\omega)Ue^{j\omega t}$.

By way of example, given a process model in a differential equation format such as:

$$\frac{d^2 y}{dt^2} + 2\frac{dy}{dt} + 10y = \frac{du}{dt} + 5u \quad (5.12)$$

assuming zero initial conditions, the application of the Laplace transform results in

$$(s^2 + 2s + 10)y(s) = (s+5)u(s); \quad \text{or} \quad G(s) = \frac{s+5}{s^2 + 2s + 10}$$

---

[9] There is always a transient behavior which vanishes after a while, if the system is stable. The concept of stability is discussed in Chap. 6.
[10] Almost any signal (Chap. 4) can be expressed by a linear combination of sine waves.

5.6.2 A Block Diagram Calculus for Linear Systems

The main advantage of the I/O operators for linear time invariant systems is the ease with which we can treat the construction of complex systems, and deal with block diagrams. It all stems from the fact that linear systems retain the spectral content of the input, and because of linearity we can deal with each spectral line in the input separately. The transfer function operators allow us to work with linear systems as if they were mere multipliers in the block diagram.

In this case, block diagrams are not only a useful tool to communicate system structure, they actually become a computational tool that allows one to quickly identify what the input-output relationships are, at least in operator form, and in the frequency domain. Operations on block diagrams allow us to perform rather complex calculus in an intuitive manner.

The block diagrams are mainly composed of the following structures (Fig. 5.8):

- *Series*: for two systems G_1 and G_2 in series, the overall operator is the product, $G = G_2 G_1$, Fig. 5.8a.
- *Parallel*: for two systems G_1 and G_2 in parallel, the global operator is the sum of them, $G = G_1 + G_2$, Fig. 5.8b.
- *Loop, (negative) feedback*: system G_1 with system G_2 in a negative feedback path, the global operator is (see Fig. 5.8c)

$$G = G_1/(1 + G_1 G_2) \tag{5.13}$$

These structures appear in repeated interconnections. We consider by way of example that in Fig. 5.9. The system is composed of three interconnected elements. Each element has an integrator (I) and a gain (K) in a particular configuration. Using the above derivations (see Fig. 5.7), we can easily write down the input-output operators for each element separately:

- Element 1's transfer function is $E_1(s) = (1 + K_1/s)^{-1}(1/s) = (s + K_1)^{-1}$
- Element 2's transfer function is $E_2 = K_2/s$
- Element 3's transfer function is $E_3 = (s + K_3)^{-1}(K_3)$

This leads to the block diagram as in Fig. 5.10.

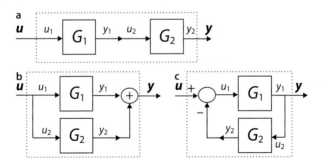

Fig. 5.8. Examples of Block Diagrams: **a** a series or cascade connection; **b** a parallel connection; **c** a negative feedback loop

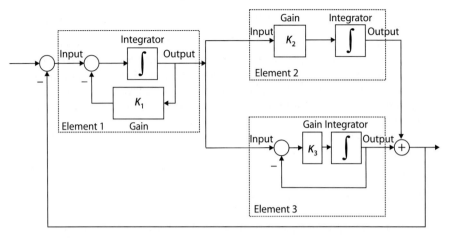

Fig. 5.9. Internal structure of a system showing the components and their interaction

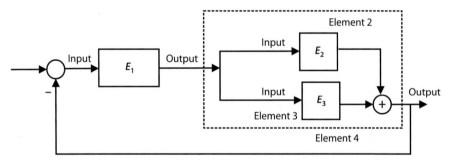

Fig. 5.10. Internal structure of the system in Fig. 5.9 after elimination of feedback loops in elements 1 and 3

Elements 2 and 3 are in parallel, their joint transfer function is the sum of both. This interconnection is called element 4. The transfer function of element 4 is $E_4 = E_2 + E_3 = K_2/s + K_3(s + K_3)^{-1}$. As indicated in Fig. 5.10 element 4 is in series with element 1. The transfer function of this series connection is their product $E_4 E_1 = (K_2/s)(s + K_1)^{-1} + K_3(s + K_3)^{-1}(s + K_1)^{-1}$. This series connection is in a feedback loop, which leads to an input-output transfer function of the form $(1 + E_4 E_1)^{-1} E_4 E_1$. After a little algebra, we find

$$G(s) = \frac{K_2(s + K_3) + K_3 s}{s(s + K_1)(s + K_3) + K_2(s + K_3) + K_3 s}$$

or after collecting terms of like power in s

$$G(s) = \frac{(K_2 + K_3)s + K_2 K_3}{s^3 + (K_1 + K_3)s^2 + (K_1 K_3 + K_2 + K_3)s + K_2 K_3}$$

Again, by putting $s = j\omega$, the frequency response is obtained.

5.7 System Analysis

Once a model has been derived for a system, we can analyze its properties. Any model will only have limited validity. The approximations that have been made in the derivation of the model are really an integral part of this model. A typical analysis will involve the evaluation of the output under various input scenarios.

When the overall system is linear, zero input analysis is important, and takes center stage. Zero input analysis provides insight into whether or not the system is well-posed. For a linear system to be well-posed all (possible) outputs from the system under zero input condition should be bounded, and preferably decay to zero. When all possible outputs under zero input condition decay to zero, the system is called *stable*. The next chapter will explore this notion in more detail.

5.7.1 Time Response

In general, understanding a system requires us to analyze the system behavior with different inputs. Typical signals (besides sine waves) that are used in systems analysis are constant inputs, or *step signal*. Alternatively, also *ramps* and *impulses* are used. Impulses are signals that apply a shock to the system, being essentially zero apart from a very short period of time during which they deliver a finite amount of energy to the system.

The output responses are typically characterized by some representative features, which are illustrated in Fig. 5.11. Most of these features are borrowed from linear systems analysis, but they are more widely used, because they allow for ready comparisons. Some of the common features found are:

- the *steady state*; the constant value the output eventually settles at, or in case of a periodic steady state, the periodic function the output eventually reaches;

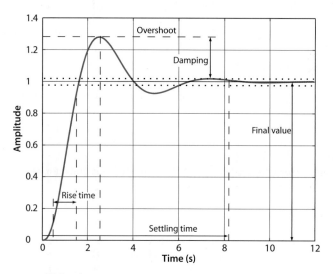

Fig. 5.11. Features qualifying the step response

- the *rise time*; the time it takes for the output to change from 10% to 90% of its final response, which is only a sensible feature provided the output reaches a constant steady state;
- the *settling time*; the time it takes the output to stay within less than 1% of its final constant steady state;
- the *maximum overshoot*; the largest deviation above the final steady state reached;
- the *undershoot*; the largest deviation in the opposite direction of the final steady state; it is not always present;
- *damping*; when the response shows oscillation, the amount of reduction in deviation from the steady state between successive peaks in the response.

Because it is not feasible to try to consider all possible inputs and their corresponding outputs, a system is studied for a class of representative inputs, building on our understanding of how to analyze inputs. From what we learned in Chap. 4 about signals this is a sensible approach.

5.7.2 Frequency Domain

Sinusoidal inputs are commonly used for the study of linear systems. In order to characterize the output due to a sinusoidal input it suffices to consider how the amplitude and phase of the output change for each pulsation. Secondly, because any signal can be arbitrarily well-approximated by a sum of sinusoids (see Sect. 4.3), this suffices to understand linear systems in a completely general setting. (Indeed recall that the response to a sum of inputs, is the sum of the responses due to the inputs separately.)

This method of analysis for linear systems is called *frequency domain* analysis. Much of the popularity of this approach in control engineering is due to the successful engineering design work of Bode[11]. Based on frequency domain ideas, he developed a successful design methodology for amplifiers, filters and feedback compensators first used in telecommunications networks. The main lasting impact of Bode's work is that he was able to characterize fundamental limits in what linear feedback loops can and cannot achieve (Doyle et al. 1992) through what are now called the Bode Integral Formulae.

By way of example, the frequency response of a gain system or an integrator system are rather trivial:

- The frequency response of a simple gain system is constant, that is, same gain for all frequencies and no phase shift at all.
- For an integrator where $G(s)|_{s=j\omega} = 1/(j\omega)$ the phase is constant and equal to -90, and the gain is inversely proportional to the frequency.

---

[11] Hendrik Bode, 24 Dec 1905–1921 June 1982, USA. He was an electrical engineer, famous for his analysis of electrical networks and feedback in particular. In recognition of his fundamental contributions, the IEEE Control Systems Society named their annual prize for outstanding contributions to control systems research the Bode Prize.

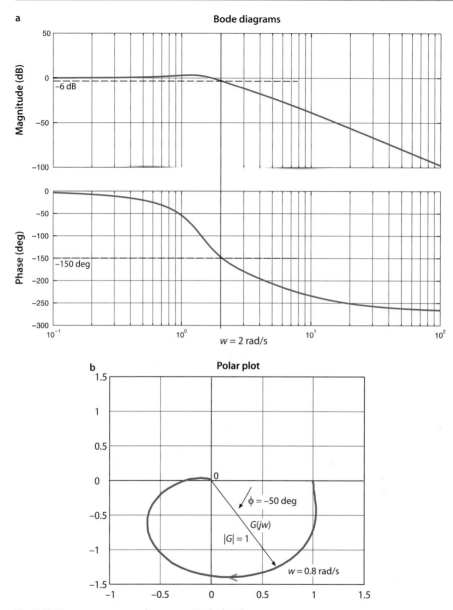

Fig. 5.12. Frequency response diagrams: **a** Bode; **b** polar

Another reason why the frequency response analysis became so popular is that a frequency response was readily graphically represented, and made calculations easy. Today, these aspects have only educational value.

Bode and/or Nyquist plots of the frequency response (Eq. 5.10) represent the amplitude and phase shift changes of such systems. The Nyquist plot (Fig. 5.12b) is simply a polar representation of the frequency response, the collection of points $G(j\omega)$ in the complex plane for all real ω. In the figure, the gain $|G(j\omega)|$ and phase $\phi(j\omega)$ for a specific frequency ω is marked. A Bode plot (Fig. 5.12a) represents the frequency domain operator in gain and phase. The gain is plotted in decibels $20\log_{10}|G(j\omega)|$ against the frequency in logarithmic scale $\log_{10}\omega$. Separately, the phase shift $\angle G(j\omega)$ is plotted against $\log_{10}\omega$. In Fig. 5.12, you can see that there is a maximum gain, a *resonance peak* for a pulsation between 1 and 2 rad/s . Also, the gain is larger than 1 until a pulsation slightly over 2 rad/s. This is the system *bandwidth* (the output has less than half the energy of the input for signals outside the bandwidth). Signals whose frequency content is well above 2 rad/s are almost completely suppressed by the system: the system is *low pass*.

In the context of nonlinear systems, harmonic analysis, that is the use of sinusoidal inputs is still of great value. Of course, because the system is not linear the output will contain harmonics, possibly even sub-harmonics. Moreover the decomposition of inputs into linear combinations of well understood inputs is no longer useful.

Example: Lightly Damped Robot Arm

Consider a highly simplified model for a lightly damped robot arm used in space explorations. The input is the torque and the output is the position of the arm. The Bode plot is presented in Fig. 5.13.

One of the advantages of Bode plots is that pictures like these can be experimentally obtained without the need for extensive modeling. Using a signal generator for the input signal and varying the input frequency, whilst recording the output a Bode plot can quickly be gathered. Spectrum analyzers do precisely this.

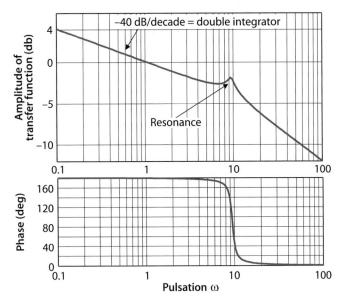

Fig. 5.13. Bode plot of the transfer function: Torque to position response for a lightly damped arm

The peak in the amplitude response is indicative of mechanical resonance. The initial slope, roughly −40 db per decade, with a phase response close to 180 degrees is indicative of two integrators in series (90 degrees phase lag per integrator, and −20 db per decade per integrator). This is to be expected as from Newton's equations we know that acceleration is proportional to torque, or position is torque integrated twice. The resonance adds a further 180 degrees phase lag in the limit, which explains why the phase starts at approximately 180 degrees and decreasing towards 0 as the pulsation increases.

5.7.3 Cascade Connected Systems

One of the nice properties of the systems analysis in the frequency domain is the simplicity to deal with cascade connected systems. In fact, we have seen that a number of elements connected in cascade are equivalent to a system, with a transfer function determined by multiplying the transfer functions of the systems in the cascade.

In analyzing the behavior of a loop, all the elements are connected in series. Thus, the gain (in decibels) and the phase of the global system will be the product of the gains (or their sum in decibels) and the total phase will be the sum of the phases of the components.

5.7.4 Integrators in Series with Feedback

In the previous example of an integrator in a loop (Fig. 5.7), if positive feedback were used (making $K < 0$), the response goes horribly wrong. Intuitively, the integrator accumulates the input that by virtue of positive feedback grows even larger, leading to an even larger accumulator and so on. This is an *unstable* behavior (to be discussed in Chap. 6). In the physical world something has to give way in the end, and typically the accumulator saturates (overflows, blows up or something similar). The problem is illustrated in Fig. 5.14a.

A single integrator in a negative feedback loop is always well-behaved, as seen in the same figure, and the Bode plot of a system composed of an integrator in a unity negative feedback loop is presented in Fig. 5.14b. The gain at low frequencies is unity (0 dB), and over the cut-off frequency (pulsation) gain decreases with almost a slope of −20 dB/decade.

Two integrators in a negative feedback loop produces spontaneous oscillations. That is without input excitation (see Fig. 2.13, without friction) the output will oscillate as soon as one of the integrators had a non-zero initial condition (the ball is down or in movement) and it is known as a resonant system. When the input has a frequency which matches this frequency the output will grow indefinitely. Good approximations for such systems are used to create large amplification for little input effort. They are the basis of all our clocks.

Three integrators in series in a feedback (positive or negative) loop always produces an oscillatory and exponentially growing, unstable response.

These examples show some of the difficulties that may arise when interconnecting systems. We will come back to these when we discuss the notion of stability and sensitivity in more detail. In general, the frequency domain ideas exposed above, work well when the system is stable.

Clearly feedback has to be treated with care.

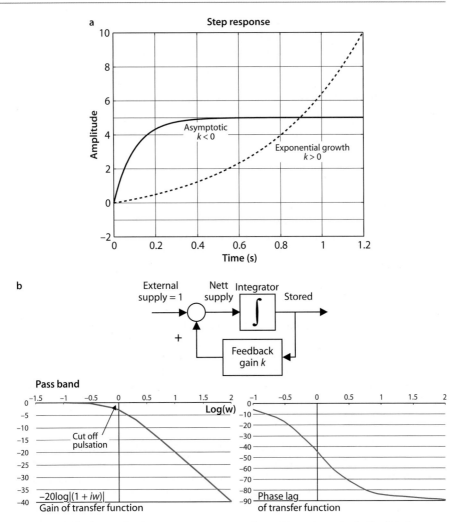

Fig. 5.14. Feedback around an integrator: **a** step response for positive and negative feedback; **b** Frequency response for negative feedback

5.8 Synthesis of Linear Systems

So far we have considered tools for the analysis of linear systems. Given a system, we may compute its response characteristics in the form of both its time response to a given input or the amplitude and phase shift as a function of the frequency. What about, given a complex model, is it possible to find an easy combination of elementary components to reproduce the global system? Or, given a desired behavior, is there a linear system with these properties?

For example, we discussed the need for anti-aliasing filters in Chap. 4. With the above ideas it is quite obvious what the characteristics of such a filter should be. Suppose we

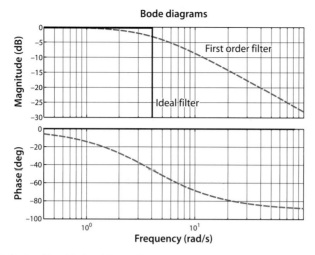

Fig. 5.15. Anti-aliasing filter, ideal and first order

sample with period h or frequency $1/h$. According to the sampling criterion the signal to be sampled should not contain any spectrum above half the sampling frequency. Hence the ideal anti-aliasing filter (Fig. 5.15) is a linear system with amplitude response $G_A(\omega) = 1$ for $\omega < \pi/h$ and $G_A(\omega) = 0$ for $\omega > \pi/h$ and preferably $G_\phi(\omega) = 0$ for $\omega < \pi/h$. It turns out that we cannot actually build a linear system with these properties. The problem is that the specified transfer function is not a rational function. Although it cannot be realized exactly, it is possible to construct a rational approximation which will do the job for all practical purposes. A very simple (and common) approximation is a first order filter. In Fig. 5.15 the Bode diagram of a low-pass filter $(G(s) = \pi/(s + \pi))$ is depicted.

If a transfer function can be written as a *proper rational* transfer function[12] of the argument s (or $j\omega$) it can be constructed by interconnecting pure gain and pure integrator systems, as shown in the previous examples. It follows from the fundamental theorem of algebra, which states that any polynomial with real coefficients can be factored into first order and/or second order polynomials with real coefficients. Elements in Fig. 5.9 are first order.

Figure 5.16 represents a second order element and it appears clear that, by appropriately connecting first or second order elements, any rational transfer function, or its equivalent frequency response, with real coefficients can be realized. This is by no means the only way in which a rational transfer function can be realized. Many different techniques exist, and there are as many different realizations as there are ways of expressing a rational function. Nevertheless a very interesting observation may be deduced from the presently suggested approach. A rational transfer function with a denominator of degree n requires at least n integrators. It is clear that

---

[12] A rational function is proper if the degree of the numerator is not larger than the degree of the denominator. The degree of a polynomial is the largest exponent in its expression.

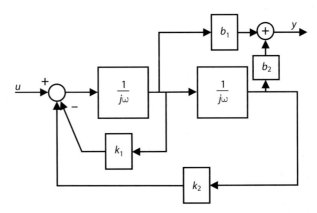

Fig. 5.16. A general realization for a second order rational (transfer function) frequency response system $G(j\omega) = (b_2 + b_1\,(j\omega))/((j\omega)^2 + k_1(j\omega) + k_2)$

we cannot do with less, and the suggested method provides a way of doing it with exactly n. So, at least the proposed method provides in some sense a minimal realization, in such a way that a minimum number of integrators are required to realize the transfer function.

There is much more to synthesis of course. The study of which integrator/gain network synthesizes a particular transfer function is known as *realization* theory. It plays an important role in electrical circuit theory. The synthesis or realization problem is particularly hard when constraints are imposed on the various building blocks that can be used to realize the transfer function. Constraints may involve limits on the gain elements (for example only positive scalars are allowed), or energy consumption limits, or accuracy limits (for example discrete building blocks with given parameters that belong to an interval). The problem of synthesis is then to construct a network consisting of fairly inaccurate (cheap) building blocks that nevertheless achieves the desired transfer function with a high degree of accuracy. There are still many open problems in this context.

The second synthesis problem, that is, to select a model (a transfer function) to match some desired specifications, is usually denoted as a design problem and it will be treated in the corresponding chapters.

The real problem in control design may be appreciated from the simple observation that control is all about designing an inverse operator. Indeed if we desire the output y to track a particular signal r, we need to find an input u so that $y = Gu = r$. In frequency domain, G is a mere multiplication, and it follows that $u = G^{-1}r$ is the input we are looking for. The problem with this observation is that normally G^{-1} does not exist as a realizable transfer function. The first reason is that if G is strictly causal (which is to be expected), then G^{-1} is not strictly causal, and obviously we cannot construct non-causal systems using causal components. In other words, the control problem is all about finding a good (simple) approximation, in the class of realizable transfer functions, for the expression $G^{-1}r$. This is known as an inverse problem. How well this can be done, depends as much on the class of signals r we need to track as on the properties of the transfer function G.

5.9 State Space Description of Linear Systems

So far we have considered linear time invariant systems and have essentially stressed the input-output relationship. Under the assumption that the system was also stable the input-output relationship is captured by the transfer function. This point of view is very compact and hides the details of the inner workings of the system.

Nevertheless, when writing down equations for systems it is almost inevitable to introduce and use variables other than the inputs and outputs of interest. Consider for example the block diagram in Fig. 5.9, where initial conditions were considered null. Let us denote by x_i for $i = 1, 2, 3$, the variables at the output of the integrator, for the respective element. The equations capturing the block diagram are most conveniently written down if we introduce also the input of the integrator, which are the corresponding derivatives of the outputs. The equations are then

$$\dot{x}_1 = -k_1 x_1 - x_2 - x_3 + r$$
$$\dot{x}_2 = k_2 x_1$$
$$\dot{x}_3 = k_3(x_1 - x_3)$$

These equations describe the integrators, the gains, the summing junctions and the interconnections. Moreover, the output of the system is

$$y = x_2 + x_3$$

These equations describe the entire system's behavior not only the input-output behavior captured by the transfer function. Moreover, the equations are suitable for computer simulation. The variable set $\{x\}$ is known as the *state* of the system. From these equations, we can of course recover the transfer function description. It suffices to replace the derivative by the variable s (or $j\omega$) and eliminate the state variables, to arrive at $G(s)$.

This process of writing down equations, eliminating variables, and finding special canonical forms for the equations that capture the entire behavior is studied in detail in the so-called behavior theory (Polderman and Willems 1998).

More generally, the set of n linear differential equations above can be written in a matrix and in compact form:

$$\dot{x} = Ax + Bu$$
$$y = Cx + Du \tag{5.14}$$

The variable x is the column vector $(x_1, x_2, ..., x_n)$, which is the state; A is a coefficient matrix, the so-called *system matrix*; B is a column vector $(b_1, ..., b_n)$, the so-called the *input-gain* vector; C is a row vector, the *output-gain* matrix; D is a scalar signifying the *feedthrough* gain. The system matrix captures the internal structure of the system and determines many fundamental properties (see next chapter), whereas the input-gain and output-gain matrices can be modified by adding, modifying or deleting some actuators (to control) or sensors (to measure) from the process. The direct coupling (D)

is most of the time null. From the input/output viewpoint, the set of matrices (A, B, C, D) is equivalent to the transfer function[13].

By means of the state space model the full system behavior can be computed. For $u(t) = 0$ and some initial state $x_0 = x(t_0)$, the system time evolution can be computed, being denoted as the system's *free response*. This is especially interesting for dealing with autonomous systems. On the other hand, if the initial conditions vanish, the system response to a given input, *the forced response*, can be evaluated. This is precisely the system output we can get from the transfer function.

The state variables possess a number of interesting properties:

- Memory. The state summarizes the past. Given the present state and the future input the future state can be computed.
- Dynamics. The effect of the input is directly connected to the derivative (the change) in the state vector. It follows that the current value of the state does not depend on the current value of the input (nor can it by definition of state).
- Completeness. From the knowledge of the current value of the state vector and the present and future input any other internal variable can be computed.
- Not unique. The state representation is not unique. Given a state representation one can derive an entire family of equivalent representations using coordinate transformations (different labels for the same space), or add more state variables (any linear combination of a state vector can be added to the state) without changing the system's input/output description.
- Minimality. There is a state vector of minimal dimension, and all other representations must have larger dimension.

Often it is possible to assign simple physical meaning to the state vector. In any electro-mechanical system it is possible to assign a state by labeling all energy reservoirs. In mechanics a state is assigned to either potential or kinetic energy, in electrical systems we assign a state for each reservoir of the electric field's and magnetic field's energy. Referring to the oscillating ball in Fig. 2.13, its behavior can be fully determined if the ball's position and speed are known at any given instant of time. These two variables define the state vector. Note that these variables represent the kinetic and potential energy. This property will be also used to analyze the system behavior. In many other instances, like in describing the world economic system, a state may have no real physical meaning at all.

5.10 A Few Words about Discrete Time Systems

Nowadays, models, simulated systems, controllers and many other systems' implementations rely on digital systems.

More than describing the possible special techniques available for the modeling, analysis and design of discrete time systems, a summary is now presented following the main concepts already developed for continuous time systems.

---

[13] Using matrix algebra, the transfer function can be directly computed from $G(s) = C(sI - A)^{-1}B + D$.

Discrete Transfer Function

The system variables are only considered at discrete time instants, conveniently represented as (positive) integers $k = 0, 1, 2, \ldots,$. Thus, the models relate the different variables at these time instants.

- A constant gain is as before, $y(k) = Ku(k)$. The input output operator or transfer function is also K.
- A unit delay (a delay of one sample period) is clearly expressed by $y(k) = u(k-1)$. In Z-transform notation it yields: $y(z) = z^{-1}u(z)$. The delay transfer function is

$$G(z) = z^{-1}$$

- An accumulator (integrator) will be expressed by $y(k) = y(0) + \sum_{j=0}^{k} u(j)$. Thus, the transfer function will be

$$y(k) = y(k-1) + u(k) \Rightarrow G(z) = \frac{1}{1-z^{-1}} \qquad (5.15)$$

Based on these elemental models, the transfer function of any linear system can be expressed as a rational function of z.

State Space Representation

In a similar way, the internal model of a discrete time system is like Eq. 5.14

$$\begin{aligned} x_{k+1} &= Ax_k + Bu_k, \\ y_k &= Cx_k + Du_k. \end{aligned} \qquad (5.16)$$

These equations are easily translated into computer programs. In a generic programming language, the code to implement is something like:

```
repeat each period
   get the new input value uk
   update the state: xk <=A*xk+B*uk
   compute the output: yk=C*xk+D*uk
   delay until next_period
```

5.11 Nonlinear Models

When dealing with most practical systems, nonlinearities always appear. For chemical or bio-chemical processes linearity is a white raven. The foundation of most chemical interactions is the so-called mass-action principle (Lotka 1998; Roberts 2009) which is inherently nonlinear[14].

---

[14] When two reactants, A and B, react together at a given temperature the chemical force between them is proportional to the active masses, $[A]$ and $[B]$, each raised to a particular power ($\alpha [A]^a [B]^b$).

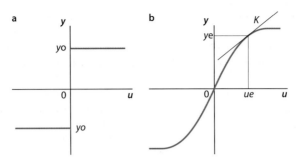

Fig. 5.17. Nonlinear static gain: **a** relay; **b** saturation

Expressed in discrete time, the state space description of a nonlinear system often takes the form,

state equation: $x(k+1) = f(x(k), u(k), k)$, $x(0) = x_0$ (5.17)
output equation: $y(k) = g(x(k), u(k), k)$

where the functions f and g are the state transition map and the output map respectively. The initial value is given at an initial time $k = 0$ as x_0. Time is represented by the integer k.

Similarly in continuous time, the general state space description (Eq. 5.14) takes the form (with $t \geq 0$ a real variable):

state equation: $\dot{x}(t) = f(x(t), u(t), t)$, $x(t_0) = x_0$ (5.18)
output equation: $y(t) = g(x(t), u(t), t)$

We can distinguish between hard and soft nonlinearities. If we consider just static nonlinearities, the idea is illustrated as in Fig. 5.17. For a hard nonlinearity, no linear approximation makes sense. The behavior of a relay presents a hard nonlinearity. For a soft nonlinearity, as in the case of saturation, a linear approximation can be useful for a selection of signals around an operating point ((ue, ye) in the figure).

With the entry of nonlinear phenomena, all of the simplicity of linear systems disappears. The result of a sum of actions is no longer the sum of the resulting effects. As such, the mathematical treatment of nonlinear systems is much more complicated. On the other hand, nonlinearities are unavoidable and they allow for unique dynamic behaviors. Hence they may be introduced by design, just like in biology. The typical approach is to somehow consider a nonlinear behavior as consisting of many local behaviors (small signals around signals of interest) for which linear techniques can be used. This idea is the cornerstone of all simulation of nonlinear systems as well as their analysis.

5.12 Comments and Further Reading

The properties of linear systems are well-studied. There is much literature dealing with all aspects of analysis, computational representation, complexity and also synthesis.

The input-output, state-space and more recently the behavioral framework have received a lot of attention: Wonham (1979), Brockett (1970), Kailath (1980), Kučera (1979), Polderman and Willems (1998), to cite but a very few. Computational methods to deal with very large linear systems are well-established, for example Antoulas (2005).

For nonlinear systems the picture is far from complete, and this will remain so, as truly nonlinearity is *the absence of definition* (to quote Prof R.R. Bitmead). From a dynamical systems point of view the following provide an introduction Wiggins (2003), Guckenheimer and Holmes (1986) and Mees (1991). From a control systems point of view nonlinear systems are treated in Isidori (1995), Khalil (2002) and Nijmeijer and van der Schaft (1990).

The question of how to find a model is well-studied, as in general this question is not very well-posed, there remains much research to be done. We borrowed liberally from the language of *behaviors* to discuss the main issues of modeling. This development is due to Jan Willems for example Willems (1970). How to obtain models from data is also a classic topic in *econometrics*, as there are few physical principles in economics. In systems engineering, the same goes under the name of *system identification*. For linear system identification, a time domain perspective can be found in Ljung (1999), and a frequency domain perspective in Schoukens and Pintelon (1991), an econometrics and more statistics-based approach is expounded in Hannan (1967).

There are many simulation languages and simulation engines to represent and compute the behavior of models. When dealing with truly large complex systems, simulation engines are typically restricted to a particular domain, as domain knowledge is where the intellectual property can most easily be protected. For example, SPICE[15] developed at the University of California Berkeley's electrical engineering department is an electric circuit simulator. CADENCE^{TM} [16] an off-spring of this development does the same for very large scale integrated circuits, providing both electrical as well as thermodynamic modeling of circuits, from design to final chip. ANSYS POLYFLOW^{TM} [17] provides for modeling of polymer processing in complex geometries. General chemical process engineering modeling is provided by AspenTech^{TM} [18]. There are many developments along these lines. Generic system simulators can be found in languages such Matlab^{TM} [19], or Scilab or LabVIEW^{TM} [20]. As systems biology, and modeling of biological systems is becoming important in developing engineered systems, systems approaches for biological models are under development (see e.g. *http://www.sys-bio.org/*).

---

[15] For more information consider the www-pages at *http://bwrc.eecs.berkeley.edu/Classes/IcBook/spice*.
[16] From C(-programming) to silicon, is their slogan, see *www.cadence.com*.
[17] ANSYS provides finite element modeling products, *www.ansys.com*.
[18] *www.aspentech.com*.
[19] MathWorks.
[20] National Instruments.

Chapter 6
Stability, Sensitivity and Robustness

6.1 Introduction and Motivation
6.2 Some Examples
6.3 Stability of Autonomous Systems
6.4 Linear Autonomous Systems
6.5 Nonlinear Systems: Lyapunov Stability
6.6 Non-Autonomous Systems
6.7 Beyond Equilibria
6.8 Sensitivity
6.9 Comments and Further Reading

Chapter 6

Stability, Sensitivity and Robustness

> *Stability is desirable,*
> *Sensitivity is a sign of responsiveness,*
> *Robustness provides a feeling of safety*
> *but great stability, low sensitivity and extreme robustness*
> **will make for a boring life!**

6.1 Introduction and Motivation

We are interested in the dynamic behavior of systems, more particularly we are interested in changing the behavior of a system, but before we attempt to synthesize desired behavior it pays to analyze system properties.

One way that changes can be effected is by tuning some parameters, like changing a gain. Another more interesting mechanism through which we may change the overall behavior is by interconnecting the system of interest with another system, which we will often refer to as a control system; the interconnected system becomes a controlled system. Our analysis will need to cater for both mechanisms.

This chapter is devoted to the notions of *stability, sensitivity* and *robustness*. Stability plays an important role in the study of system dynamics. It captures the idea that *small causes have small consequences forever.* This is obviously a desirable property in particular when we want to predict future behavior of a system with a degree of certainty and accuracy. All *controlled* systems should be stable.

Of course, our notion of small has to be made precise, and how we determine the size of signals and systems plays an important role in defining what we mean by stability. Also we may expect many different notions of stability to distinguish between the different types of cause and effect we are interested in. As usual, we will only consider a few of the options.

Most important to the notion of stability is the fact that it relates to how the system evolves over time into the indefinite future. Simply requiring that small causes have small effects is called *continuity,* and more often than not we will take this property for granted.

Stability is first a qualitative property, in design however it is important to be able to quantify how stable a system is. A measure of stability, like the range of parameters we can accept in a system without losing stability, is critically important in any design problem. To this end, in systems' analysis the notions of sensitivity and robustness are used.

Sensitivity measures how important signals, like the input and output of a system, depend on other external signals affecting the system, like measurement noise or actuator disturbances. If a small noise signal can change the input to the plant significantly, we say that the input is very sensitive to this noise, a highly undesirable situation.

Robustness indicates how changes in the parameters or the environment of a system may affect the system behavior. Typically we like the response of a system not to vary too much despite the fact that the system may undergo some changes.

Most natural and engineered systems behave in a stable manner and do not rely on control for the purpose of stability. There are important exceptions though and some examples are presented in the next section. In any case, control will be called for if the measure of stability, as also reflected in sensitivity and robustness properties, does not meet expectations.

This chapter is organized as follows. First we look at some simple examples, to motivate and to set the scene somewhat. Then we look at the notion of stability, from its mechanical origins' point of view, for systems without inputs, so-called *autonomous systems*. From a control point of view systems without inputs are rather boring, but they form a good place to start. Next we provide some more detail for linear systems. Then we consider stability from a general system dynamics perspective, particularly dealing with inputs. The ideas of robustness and sensitivity come natural within this context.

6.2 Some Examples

By way of setting the scene, consider a direct current electrical motor, as shown in Fig. 6.1.

Whenever we apply a voltage to the armature, E_a, after some *transient* period, the motor will reach a stable shaft speed, ω. For each applied voltage, a particular motor torque results, P_m. Depending on the mechanical load, L, a particular final speed is attained. From this point of view the motor behaves in a stable manner.

One notion of sensitivity captures how big a voltage change is required to effect a particular change in rotation speed. Sensitivity is high when small changes in the input voltage produce corresponding and large variations in the output speed. In the same context a notion of robustness captures how big a change in speed is observed for a particular change in mechanical load on the motor axle. More robustness would mean that a large change in mechanical load only results in a minor variation of the speed, a very desirable property in many applications. This would be the case for a highly geared motor, where the motor speed is much higher than the output shaft speed (see for example Sect. 3.4).

Any description of stability, sensitivity and robustness in this context would only apply within what are called the operational limits for the electrical motor. For example due to *saturation* effects in the magnetic material used in the motor construction there exists a voltage level beyond which a further increase no longer yields an

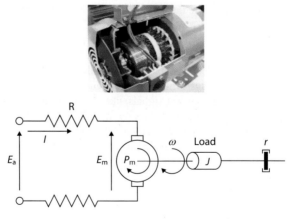

Fig. 6.1. A DC motor

increase in motor speed. Even worse, there is a voltage beyond which the motor will be irreparably damaged, and start to smell unpleasantly. Normally system models for control purposes will only describe the system of interest, here the electrical motor, for signals well within these operational limits. In other words, the model's validity (and consequently the usefulness of any notions such as stability, sensitivity and robustness) will break down well before the system does.

Not all systems are stable. There are inherently unstable systems. A familiar example of a mechanically unstable system that needs careful control to function properly is our body. We need control to keep our balance and our posture whilst walking, standing even sitting. In order to walk, we all must learn how to keep our body, in particular our head, in an upright position above our feet. Mechanically (in the presence of gravity) our upright position is somewhat unnatural, it is an *unstable* position. On all fours we are rather more stable, but far less mobile. Mobility and stability always require a trade-off.

A similar situation, not so easily managed, is to balance a slim long rod upright in the palm of your hand, as the general picture of an inverted pendulum (Fig. 6.2)[1] shows. Exactly the same problem occurs when launching a rocket for say a space shuttle. The long rocket body must be balanced against gravity using a thrust force at the bottom of the rocket. This requires careful control of the orientation of this thrust force.

In some chemical processes an unstable equilibrium (that is without control it is unstable) is desired because the yield is much higher at this equilibrium as compared to the possible naturally stable equilibria. Similarly, in a mechanical system, an instability (in the uncontrolled system of course) may lead to high maneuverability and hence be desirable.

The consequences of letting an unstable process evolve unchecked can be disastrous, as is well-known to us from the Chernobyl nuclear power plant's catastrophic accident. Graphite moderated nuclear fission reactions are unstable at low power levels, and stable at high power output, where they should be operated. The lack of proper control and sufficient fail safe mechanisms in the low power regime in the Chernobyl nuclear reactor caused a massive explosion in April 1986.

Fig. 6.2. An inverted pendulum

---

[1] Taken from Quanser®.

6.3 Stability of Autonomous Systems

The concept of stability was first coined in mechanics, which is the study of motion of objects in three-dimensional space (like robotics or the motion of satellites).

Stability is a local property attached to a particular response (motion) in a system. What happens to the system response when at a particular instance in time a small deviation (in position or velocity) is introduced? Is the subsequent evolution close to the unperturbed response? If it is, we say the response (motion) is stable, if not the response (motion) is unstable.

More often than not we discuss this property only for very specific system responses. Sometimes, when there is no room for confusion, we may say that the system is stable, meaning that any response in the system has this property.

Consider the example of the motion of a ball in some undulating terrain, subject to gravity and rolling friction. The situation is somewhat abstractly represented in Fig. 6.3. The *system* here is the ball rolling over the surface subject to gravity. If we assume that the ball is always in contact with the surface, it is fully characterized by its longitudinal position and speed.

Intuitively it is clear that the lowest position in the valley, **a** in Fig. 6.3 is a special point. It is called an *equilibrium point*. If the ball is there initially at rest (that is without velocity), and no external force, other than gravity and friction are applied (which implies that the ball's acceleration in the horizontal direction is zero), it will remain there forever as predicted by Newton's laws. A small perturbation away from this point (either in position, or by giving a small push, or by providing an initial horizontal velocity) is not going to cause much concern, the ball will remain in the neighborhood of the equilibrium **a**, or its motion will be close to the equilibrium at all times in the future. This equilibrium is called *stable*.

The position **c**, on top of a hill, is also an equilibrium point but it is rather different from the equilibrium **a** in the valley. A small perturbation (either by a non-zero horizontal velocity or a small horizontal force) away from the position **c** will result in the ball running down, away from **c**. Moreover the ball will not return back to **c**. We capture this, by saying that the equilibrium **c** is *unstable*.

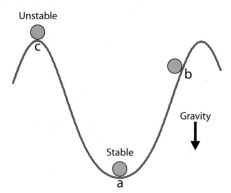

Fig. 6.3. A ball, subject to gravity (assumed directed from top to bottom of the page), on an undulating surface

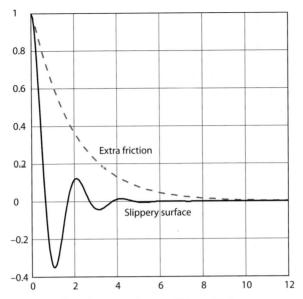

Fig. 6.4. Trajectories starting from the same point, with different friction

When the ball is released from an initial position **b**, with zero initial velocity, the ball will roll down under the influence of gravity and start oscillating back and forth around the position **a**. Because of friction the ball will eventually settle at the point **a**. We summarize this as the point **a** is a *stable and attracting* equilibrium point. It is also clear that the equilibrium **a** is only locally stable. Indeed if we give the ball a sufficiently large initial velocity it will be able to pass the position **c**, never to return.

The trajectory of the ball, projected on the x-axis, that is, on the horizontal, will strongly depend on the friction coefficient between the ball and the surface it rolls over. The trajectory may be oscillatory, if there is little friction to damp the oscillations, or without oscillations if the ball is loosing much energy as it moves, as pointed out in Fig. 6.4.

Stability (instability) is a local property, a qualitative descriptor for the behavior of the motion of the ball near a particular motion (here we used equilibria). It considers the motion of the ball in the future, so from the time point of view it is a global property, that is what makes it so special. That stability is a local in the motion space is quite clear from the example, as the ball has multiple equilibria with different stability properties.

The stability property that the equilibrium **a** possesses is very strong. First, we may perturb the starting point, or provide a small initial velocity, or a small push to any trajectory in a neighborhood of **a**, and any such generated trajectory or motion will remain close to **a** and eventually will converge to **a**. It is said that the equilibrium is *asymptotically stable*. Asymptotic stability requires that the motion is not only stable, but that is also attractive, nearby motions converge to the motion under consideration.

> **Stability of an Autonomous System**
>
> Consider a dynamic system as described by the collection of responses from all possible initial conditions.
>
> An equilibrium is a constant response.
>
> We say that the equilibrium is *stable* if for any initial condition close to the equilibrium, the corresponding system trajectory remains close to it for all future time. Otherwise, we say that the equilibrium is unstable.
>
> We say that the equilibrium is *attractive* if for any initial condition close to it the corresponding system trajectory converges to it as time progresses.
>
> We say that the equilibrium is *asymptotically stable* if it is both stable and attractive.
>
> We say that the properties hold globally if there is no restriction on the initial condition.
>
> A *trajectory* (system response) is *stable* if small initial deviations away lead to small deviations in the indefinite future. It is asymptotically stable if it is stable with the added property that the future deviations become vanishingly small.

6.4 Linear Autonomous Systems

In the case the dynamics are linear, the question of stability can be settled using linear algebra. Most stability questions, including synthesis questions, are computationally tractable, even if the state dimension is large. Substantial computer-aided design software tools are available.

Though few systems are linear, most systems allow for a linear system that approximates it well enough over the operational range of interest to test for stability even in the nonlinear situation. After all stability is a local property. This is in a nutshell the content of the famous Hartman-Grobman theorem in dynamical systems. Moreover, in a control context, the whole aim is typically to model for control, and most likely control requires that the controlled system exhibits a desirable behavior. It is clear that good performance would mean that the controlled system should be close to this desired behavior. The deviations away from this desired behavior can often be well approximated by a linear system.

Consider the simplest autonomous dynamic system model

$$x(k+1) = ax(k), \quad k = 0, 1, 2, \ldots \tag{6.1}$$

where $x(k)$ and a are scalars. a is called the **pole** or also *eigenvalue* of the System 6.1. It is easy to compute the collection of all future solutions, by simply iterating (6.1):

$$\{x(k) = x_0 a^k \quad k = 0, 1, 2, \ldots\}$$

where x_0 is the initial condition $x(0) = x_0$.

It is also clear that $x_0 = 0$ is an equilibrium.

Clearly this equilibrium is globally asymptotically stable provided $|a| < 1$, the pole must be less than one in magnitude. Indeed, under this condition all solution stay close to the equilibrium and converge to it as time progresses.

With $a = 1$ or $a = -1$ the equilibrium is stable, but not attractive. For $a = 1$ all solutions remain constant, so what starts close remains close, but clearly no solution gets closer to another. For $a = -1$ all solutions, apart from the equilibrium are two-periodic: $x(0) = x_0$, $x(1) = -x_0$ and $x(2) = x_0$ and so on; clearly the equilibrium is stable but solutions are not attracted towards the origin.

If $a > 1$ or $a < -1$, the equilibrium is unstable, and all solutions (apart from the solution that stays at the equilibrium) diverge.

Using the Z-transform, the System 6.1 can be rewritten as

$$zX(z) - x_0 = X(z), \quad \text{or} \quad X(z) = \frac{1}{z-a} x_0 \tag{6.2}$$

The system operator[2] $1/(z-a)$ captures the stability of the system.

6.4.1 General Linear Autonomous Systems, Discrete Time[3]

General linear systems in discrete time can be decomposed in a set of first order equations, like Eq. 6.1, with variables x_i for $i = 1, 2, \ldots, n$. Obviously, in this case the system will have n poles or eigenvalues.

Another model for a general linear discrete time system, without input, as introduced in Eq. 5.16 is

$$x(k+1) = Ax(k), \quad k = 1, 2, \ldots \quad x(0) = x_0 \tag{6.3}$$

where $x(k)$ is the state vector at time k, and A is called the system matrix, or state transition matrix.

The stability of the equilibrium $x = 0$ can be settled by considering the n eigenvalues of the matrix A.[4] It can be established that stability is guaranteed when all eigenvalues of the matrix A have a magnitude less than 1.

This is illustrated in Fig. 6.5, which shows a number of eigenvalue locations, with the associated time functions, clearly delineating stable and unstable behavior.

Another representation of a linear system is through the linear difference equation:

$$y(k+n) + \alpha_1 y(k+n-1) + \ldots + \alpha_n y(k) = 0 \tag{6.4}$$

or, by means of the Z-transform

$$(z^n + \alpha_1 z^{n-1} + \ldots + \alpha_n) y(z) = 0 \tag{6.5}$$

the stability is determined by the roots, a_i, also called *poles*, of the *characteristic equation*, namely

$$z^n + \alpha_1 z^{n-1} + \ldots + \alpha_n = \prod_{i=1}^{n}(z - a_i) = 0 \tag{6.6}$$

---

[2] The Z-transform of the sequence $x(k)$ is by definition given by $\sum_{k=1}^{\infty} x(k) z^{k-1}$. Using the explicit expression for the solution $x(k)$ we obtain $\sum_{k=1}^{\infty} a^{k-1} x_0 z^{k-1} = 1/(z-a) x_0$.
[3] This section may be skipped in a first reading.
[4] A scalar λ is an eigenvalue of the matrix A if there exists non zero ξ such that $\lambda \xi = A\xi$. Eigenvalues can be complex numbers. Selecting ξ as an initial condition, we obtain the solution $x(k) = \lambda^{k-1} \xi$. Clearly the equilibrium cannot be stable as soon as there is an eigenvalue such that $|\lambda| > 1$.

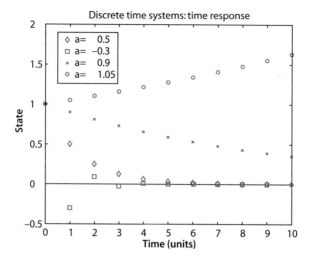

Fig. 6.5. Stability of linear systems in discrete time can be verified by the location of the eigenvalues of system matrix A

The system is stable if $|a_i| < 1, \forall i$. The system is stable if the absolute value of each pole of the system is less than one. These poles are complex numbers and, stability follows if all the poles are located inside the unit circle in the complex plane.

6.4.2 Continuous Time, Linear System

If a continuous time first order autonomous (without input) system is considered, its model, similar to Eq. 6.1, is given by

$$\frac{d}{dt}x(t) = \dot{x}(t) = ax(t), \quad x(0) = x_0 \tag{6.7}$$

This model was already introduced in Eq. 4.3, when dealing with exponential signals, and we know that the solution of this equation, the trajectories of the system from a given initial condition x_0, is

$$x(t) = e^{at} x_0$$

It is clear that $x_0 = 0$ is an equilibrium point and that in order for this equilibrium to be (globally, asymptotically) stable the pole, a, must be negative. Otherwise, the solutions will exponentially diverge.

Using the Laplace transform[5], the System 6.7 can be rewritten as

$$sX(s) - x_0 = X(s), \quad \text{or} \quad X(s) = \frac{1}{s-a} x_0 \tag{6.8}$$

Again, the system operator $1/(s-a)$ captures the stability of the system.

---

[5] See Laplace transform property (Eq. 4.12).

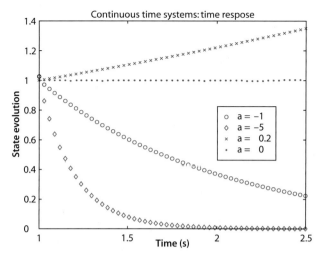

Fig. 6.6. Stability of linear systems in continuous time can be verified by the location of the eigenvalues of system matrix A

For general linear systems in continuous time, the model can be also decomposed in a set of first order equations, like Eq. 6.7, and the set of coefficients a_i, called poles or eigenvalues of the system, will determine the stability condition of the system. The s-operator in this case is a composition of terms $1/(s - a_i)$, globally expressed as a rational function which denominator is $(s - a_1)(s - a_2) \ldots (s - a_n)$, is denoted as the characteristic polynomial of the system.

The system operator captures the stability of the system as well and hence the system is stable provided the poles of the transfer operator are strictly in the left half of the complex plane (the real part is negative). This is illustrated in Fig. 6.6, which shows a number of eigenvalue locations, with the associated time functions, clearly delineating stable and unstable behavior.

6.4.3 Exploring Stability

In designing control systems the minimal property to establish is stability. It is therefore important to understand how the stability property depends on system parameters. Is stability a sensitive property? Can small variations in parameters cause system stability to change? The example of the ball in the valley indicated that this may not be the case. On the other hand, if a system is unstable, the main goal of control is to be able to change instability into stability (that is we must be able to create a valley where there used to be a crest). There must be a transition from stable to unstable somewhere as parameters vary. How this happens, and what actually happens in the behavior of a particular system as parameters in the system vary belongs to the realm of *bifurcation theory*. In great generality it can be shown that system behavior varies typically smoothly with parameter changes, apart from some singular combinations of parameters. Around such *bifurcation* points, qualitative properties such as stability may change abruptly, because equilibria are created or disappear or interact in some manner.

An Example

We go back to the example of the tank that we considered in Chap. 2, see in particular also the Sect. 2.4.4.

Consider a water tank, described as follows

$$x(k+1) = x(k) - c(k) + u(k); \quad u(k) = \alpha(x(k) - r) \tag{6.9}$$

where
- $x(k)$ is the water volume in the tank at the kth instant of observation,
- $c(k) \geq 0$ is the total volume of water removed from the tank over the period between the kth and $k+1^{st}$ observation. It is assumed to be independent of the tank volume,
- $u(k)$ is a volume of water supplied or extracted over the same period,
- $r > 0$ is the desired water volume in the tank,
- α is a *control* parameter adjusting how much water is taken from or added to the tank[6].

Assume for the moment that the volume of water taken from the tank over a sample period is constant $c(k) = c > 0$ and that the desired water volume r is also constant. We are going to analyze the behavior of the tank based on the control parameter α. How does it influence the dynamics of the system?

It will be useful to rewrite the system description in the following manner:

$$x(k+1) = (1+\alpha)x(k) - c - \alpha r \tag{6.10}$$

Equilibrium Point

First there is an equilibrium point x_e, which follows from:

$$x_e = x_e - c + \alpha(x_e - r), \quad \text{or} \quad x_e = \frac{c}{\alpha} + r \tag{6.11}$$

The physical interpretation is that, at equilibrium, the water added to or taken from the tank through the control term $u(k) = \alpha(x_e - r)$ is constant, and exactly compensates for the water extracted c. Under this condition, the tank's water volume stays constant. Notice that this immediately implies that the equilibrium water level cannot be equal to the desired level (Eq. 6.11). The control term $u(k)$, based on feedback (observation of the tank level, relative to the desired level), needs an error $x_e - r \neq 0$ in order to be able to take a corrective action.

The System 6.10 can be rewritten in yet another equivalent format:

$$(x(k+1) - x_e) = (1+\alpha)(x(k) - x_e) \tag{6.12}$$

showing the evolution of the difference between the tank volume and the equilibrium point.

---

[6] This kind of control is denoted as proportional control, as described later in Chap. 10, the volume of water added is proportional to how much the tank volume deviates from the desired volume.

Design Question

The design question is "When is the equilibrium stable?". How do we need to select the parameter α to ensure stability, as well as an equilibrium that is close to the desired level[7]?

The equilibrium x_e is to be close to the target level $r > 0$. This requires that a large value for α (either positive or negative) is the way to go. But, to be stable, the pole of the linear system, (Eq. 6.12), $1 + \alpha$ should be less than one in magnitude, which severely limits the range of acceptable α

$$0 > \alpha > -2$$

(see the previous Sect. 6.4). It follows that stability requirement implies negative feedback $\alpha < 0$. If the water fill volume is too high, i.e. $x(k) - r > 0$, water is taken from the tank by the control action $u(k)$ as $\alpha < 0$.

It is interesting to visualize the evolution of the water volume by plotting it in the plane whose horizontal axis represents the current state, and the vertical axis the next state. Equation 6.12 is handy for this purpose.

A very special case, described as *dead-beat* control, happens when $\alpha = -1$. The equilibrium is reached after the first control action.

In Fig. 6.7, the trajectory of the water level is traced for $\alpha = -1.5$. Similar graphs can be plotted for all α, changing α is reflected in the picture through a changed slope for the off-diagonal line (do that as an exercise).

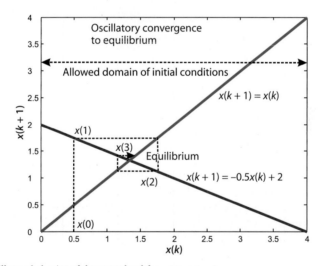

Fig. 6.7. Oscillatory behavior of the water level for $-1 > \alpha > -2$

---

[7] The fact we are asking two questions, and have only one degree of freedom to play with, immediately implies that there will have to be a trade-off between these two objectives.

Bifurcations

Most importantly the example illustrates how we change stability into instability through feedback and how this transition occurs. All of these phenomena may be summarized in a single diagram, in the parameter plane Fig. 6.8. This figure indicates for each of the possible choices of water reference level r and control parameter α a specific qualitative system behavior. Such a diagram is known as a *bifurcation* diagram. It identifies the loci in parameter space where the qualitative properties of the system (or more precisely its trajectories) change abruptly (or bifurcate).

We conclude this example with a few generalizations.

- Stability properties change at discrete parameter values (here $\alpha = 0, -1, -2$). Away from these special conditions a small change in a parameter only affects the trajectories, their stability and performance attributes in a small way. This is true in great generality. It is one of the foundational observations in so-called *bifurcation* theory, which is all about understanding these special points and organizing them so that we get a clear overview of what happens (or may happen) to the trajectories of a dynamical system.
- The dynamics in the example are linear. However, as discussed in the next section, and because of the Hartman-Grobman theorem, the results carry over to equilibria of nonlinear systems as well. In a small neighborhood of equilibria, a linear system approximation can represent most of the local behavior as well as the nonlinear system.
- Performance in steady state (here the difference between the equilibrium and the desired level) and performance in transient (here how fast the equilibrium is reached) are coupled, but they do not go hand in hand. There is a trade-off. Good steady state per-

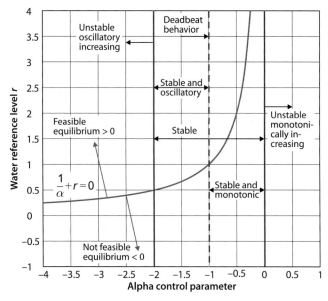

Fig. 6.8. The changes in behavior for the water level in a tank subject to a constant water flow drain $c = 1$, for variable water reference r and control parameter α

formance would require large α, but the largest α which also guarantees stability is -2. Moreover, the closer α is to -2, the slower the transients die out, or the longer it takes to get to the equilibrium. The system structure is too simple to allow us to design for transient as well as steady state performance. More elaborate *control* over the supply rate must be considered to avoid or improve on the encountered trade-off. It is precisely the existence of these trade-offs that make (control) design interesting.

6.5 Nonlinear Systems: Lyapunov Stability

The development of the notion of stability is intimately associated with the name of Lyapunov[8], a Russian mathematician interested in the stability of motion as studied in mechanics. For his master research program he was given the topic to investigate the shape a viscous liquid would assume when rotating around an axis of symmetry. This was and still is an important question in understanding the shape of planets. The problem proved rather difficult (it is still an interesting object of research) and after he had defended his Master's thesis Lyapunov abstracted, and simplified the problem to the study of the motion of rigid bodies (as opposed to deformable ones) with an arbitrary but finite number of degrees of freedom and subject to given force fields. He developed a theory, now called Lyapunov theory, that allowed one to infer stability without the need to know the precise motions of interest. The latter was extremely important as in general it is not possible to express the motions of interest in an analytically tractable form (apart from linear systems). Even today, with all the computing power we have available, Lyapunov's ideas are still relevant. Indeed stability is not easily captured through numerical simulation studies as we need to be able to verify an infinity of possibilities (for example all the possible initial positions of the ball in Fig. 6.3). His results and ideas have influenced the study of systems theory ever since, and even today they play an important role in the design and verification of nonlinear control systems.

6.5.1 Lyapunov's First Method

The first great contribution that Lyapunov made was to clearly understand that the stability of an equilibrium is a local property and that it can be studied using a linear approximation to the dynamics of interest. This is the so-called first method of Lyapunov for stability. To analyze the stability of point **a**, in the ball example, we only need to investigate the ball behavior in a neighborhood of this point. Despite the fact that stability property requires us to consider the time evolution of all trajectories in a neighborhood of the equilibrium into the indefinite future, Lyapunov showed that under very mild conditions to be imposed on the linearization, the stability properties enjoyed by the linear approximation are the same as the stability properties of the system itself. Moreover for linear dynamics Lyapunov provided a purely algebraic character-

---

[8] Aleksandr Lyapunov 1857–1918, studied mathematics in St. Petersburg University. He was influenced by L. Chebyshev to study the stability of ellipsoidal forms of rotating fluids, and obtained his Masters degree in 1884 on this topic. In 1892 he delivered his now famous Stability of Motion dissertation, on the basis of which he received his Doctorate. Another mathematician with a lunar feature named after him.

ization of stability, through the solution of a set of linear equations that now carry his name. This is a great advance, because we now have a very simple computationally efficient manner to verify the stability of equilibria.

6.5.2 Energy and Stability: Lyapunov's Second Method

The study of stability is very much linked to the notion of energy. Returning back to the example of the ball subject to gravity, we notice that the stable equilibria correspond to local minima in the potential energy for the ball, i.e. the ball at rest in a valley. Also, the local maxima in the potential energy for the ball, that is the ball at rest on a crest, correspond to unstable equilibria for the ball.

If we follow the trajectory of the ball released from point **b** with zero velocity, it is clear that the ball gains in kinetic energy (speeds up) as it loses potential energy. The potential energy that the ball possessed in point **b** is converted into kinetic energy, and some heat, a consequence of friction with the ground. When the ball reaches point **a** its potential energy achieves a local minimum, and its kinetic energy reaches a local maximum. As the ball continues to roll, it converts its kinetic energy back into potential energy, reaches a maximum position (local maximum in potential energy, local minimum in kinetic energy) and rolls backwards; and so on. Eventually the ball comes at rest at point **a**. From the first law of thermodynamics, we know that energy is conserved. Hence the mechanical energy in the ball, the sum of its potential energy and kinetic energy is decreasing over time, as friction dissipates energy in the form of heat which must come at the expense of the available mechanical energy in the ball. Moreover, as the total mechanical energy is bounded from below, because clearly the mechanical energy has a global minimum if the ball is at rest in the deepest valley, it must follow that the mechanical energy of the ball reaches a minimum along the trajectory of the ball. At that point the ball is precisely in a point like **a**, at the bottom of a valley, and possesses no kinetic energy. Natural systems evolve towards stable equilibria. (This makes a nice link with the maximum entropy principle in thermodynamics; Gyftopolous and Beretta 1991.)

For most physical systems the total energy in a system and in particular its evolution over time is always an excellent starting point to understand the system's stability. Even then, for complex systems, or system dynamics not rooted in physics, like social or economic systems, it may not be possible to think in energy terms. To overcome such problems, these energy ideas have been beautifully generalized by Lyapunov to allow one to study arbitrary abstract systems. This was Lyapunov's second important contribution to the study of the stability of motion, called *Lyapunov's second method*.

> **Lyapunov's Second Method for Stability**
>
> Consider an equilibrium point x_e of a system of interest.
>
> Suppose there is an energy-like function of the system state, $V(x)$, such that $V(x) \geq 0$ in a neighborhood of the equilibrium point x_e, and is such that it only vanishes at this point. This captures the notion that the bottom of the valley is an equilibrium.
>
> The system will be stable if, for all initial conditions x_0 in some neighborhood of x_e along a trajectory starting at x_0 the function value $V(x(t))$ (where $x(t)$ is the trajectory starting at x_0) is not increasing with time.

In Lyapunov's second method, identifying stability is now shifted to finding an energy like function V, called Lyapunov function with all the desired properties: bounded below (with a minimum at the equilibrium), and decreasing along solutions. If such a function is found then stability can be inferred.

Do such Lyapunov functions always exist when the equilibrium in question is stable? It would be important to know that a search for such a function is not in-vain. Lyapunov demonstrated that *if* an equilibrium is (locally) stable and attractive, then indeed such a Lyapunov function exists. Nevertheless, such existence results provide little solace in the quest for finding appropriate Lyapunov functions as the Lyapunov functions constructed in these existence results typically involve the collection of all trajectories of the system, which is not a computable object.

The weakness of the second or direct method of Lyapunov is exactly in finding the Lyapunov function. This is hard, and failure to find an appropriate function just means that we are unable to find it. It does not prove anything.

The strength of the Lyapunov idea lies in the fact that a conclusion about stability can be reached without precise knowledge or even being able to compute the trajectories of the system. Indeed all we need to establish is that the scalar valued Lyapunov function V) is decreasing along the evolutions of the system. It captures the behaviour of *all* trajectories (at least in a neighbourhood of the equilibrium of interest) at once, without the need of computing all trajectories. This can be established without knowing the solutions. Suppose that the system is described by an ordinary differential equation $\dot{x} = f(x)$. We may not be able to solve it, but we can check if a $V(x)$ is decreasing along the solutions, as all we have to do is to compute the derivative of V along the solutions, and this is given by

$$\frac{d}{dt}V(x) = \frac{\partial V(x)}{\partial x}f(x)$$

An example illustrates the point.

Nonlinear System Example

Consider the system defined by

$$\dot{x} = -x^3$$

A Lyapunov function candidate is

$$V(x) = x^2$$

because it is always positive, and zero at the equilibrium, moreover as $\|x\|$ grows larger so does $V(x)$. The time derivative of V along the solutions satisfies:

$$\frac{d}{dt}V(x,t) = 2x\dot{x} = -2x^4 \leq 0$$

which is always negative, except at the equilibrium. The equilibrium is stable, even asymptotically stable, and because there are no restrictions on our calculations we can even decide that the stability property holds true globally.

6.6 Non-Autonomous Systems

So far we have explored how the stability of an equilibrium, a particular solution of a system, can be determined using Lyapunov ideas, and how these stability properties change as we modify some parameters that appear in the description of the system. In this discussion all the parameters are assumed to be constants, they do not vary over time, and we observe how these constants shape the behavior of the system. In many other situations however, these *parameters* are not fixed but may vary from time instant to time instant; for example it is natural to expect that the water consumption may depend on time, as the amount of water drained from the tank will depend on the opening of a valve and the latter can change over time. Time variations complicate matters substantially but also create interesting possibilities.

Another important question in the study of stability is: *How does the interconnection of systems affect the stability of the overall system?* To discover this it does not suffice to study the stability of equilibria of autonomous systems. Interconnecting systems means sharing of signals between systems, hence it is imperative to study system stability from a signal or input perspective.

6.6.1 Linear Systems

Linearity simplifies matters considerably. Indeed, in general because in non-autonomous systems, the solutions are a consequence of both the initial conditions as well as the external inputs. We expect that in order to deal with stability in the presence of external signals we will also need to keep track of how initial conditions and external signals interact. However, in the linear system case, the solution can always be written as a linear combination of two solutions, one without initial conditions, only due to the external signal and one without external input, only due to the initial condition. That means input/output stability can be analyzed assuming null initial conditions. Moreover, using transfer operator notation, it is easy to see that the system stability does not depend on the input (Eq. 6.6).

Consider again the example of the tank. If the input variables, such as the reference r or the water consumption c, are time varying, the only result is that the equilibrium point will change, but still the system dynamics will be characterized by the parameter α.

In a more general setting we can consider the input/output behavior determined by the transfer function operator, Eq. 5.8 in the continuous time case and Eq. 5.15 in the discrete time setting[9].

---

[9] Almost any discrete time transfer function can be expressed as
$$G(z) = \sum_{i=1}^{n} \frac{b_i}{z - a_i}$$
The parameters a_i and b_i may be complex.

> **Linear Systems Stability**
>
> A linear system $x(k+1) = Ax(k) + Bu(k)$ or $\dot{x}(t) = Ax(t) + Bu(t)$ is stable if the poles (eigenvalues of A) of the System 6.6 are:
>
> - inside the unit circle (smaller than one) if the system evolves in discrete time, see Fig. 6.5,
> - strictly to the left of the imaginary axis in the complex plane (real part negative) if the system evolves in continuous time, see Fig. 6.6.
>
> Stability does not depend on the input. It is an intrinsic property of the system.

There are some computationally efficient mechanisms (using Lyapunov equations for example) to conclude whether or not the eigenvalues of a matrix lead to a stable linear system, without necessarily computing the matrix eigenvalues explicitly.

6.6.2 Nonlinear Systems

In trying to come to an understanding of how an input into a system, affects its overall behavior the main difficulty is that there is so much that can change over time. So in order to make some progress, we are going to limit the possibilities of what we want to consider. One reasonable question would be to ask how the size of an input affects the size of a signal of interest in a system. In stability literature this is captured by the notions of *input-to-state stability*, or ISS), and the concept of input-to-state gain[10] (function).

In general even this simple question is difficult. For example in the tank example, if the input is neither the reference nor the outflow but the control parameter α, the system becomes nonlinear, as there is the product of the input and the state. We may guess that, in this case, the system will become unstable as soon as the input (α) goes out of the interval $\{0, -2\}$.

To discuss the general case, we start with the discrete state space description of the System 5.17.

The initial value for the state is given at time t_0 as x_0. Recall that the nature of the state is such that given the information about the initial state and the model equation (which includes knowledge of the time variations and the input function into the indefinite future) we are in a position to compute the future behavior of the state, that is for all time after the initial time $t > t_0$.

A typical input-to-state stability property takes the form:

$$\|x(t)\| \leq g_x(\|x_0\|, t) + g_u(\|u\|), \quad \forall t > t_0 \tag{6.13}$$

Here $\|.\|$ is a measure for the size. g_x is a function both of the size of the initial condition and time. It reflects how initial condition effects disappear in time, that is we expect $g_x(., t) \to 0$ as time progresses. Also, $g_x(0, t) = 0$.

---

[10] The concept of gain is much more general than the one described in Sect. 5.5. It expresses the ratio between some output size and some input size. These signal sizes can be for example the largest absolute value reached by a signal, or its total energy. There are many different measurements of size for signals.

g_u is also a gain function, that expresses how the input affects the state. We want that $g_u(0) = 0$ and because it is a gain function, we expect that its value increases with the argument, that is $g_u(a) < g_u(b)$ whenever $0 < a < b$ and similarly $g_x(a, t) < g_x(b, t)$ whenever $a < b$. Implicitly, by stating Eq. 6.13 it is clear that there is an equilibrium at zero. So, a statement like this is a very strong statement about stability.

Clearly, ISS approaches stability both from an initial condition *and* input aspect. Indeed an appropriate ISS concept must capture an initial condition stability property, because it must capture the no input condition as a special case.

An ISS property is geared towards characterizing the stability of an equilibrium. This is the most frequently encountered stability question, but by no means the most general.

By demanding that all trajectories in some neighborhood of the equilibrium remain close and also eventually converge to this equilibrium under zero input conditions, ISS recovers the classical notion of an asymptotically stable equilibrium. In fact one can show that the previous definition and the above expression are equivalent.

In addition, ISS deals with the input and expresses that the input cannot drive the state of the system too far from the equilibrium. How large the state can become under the influence of the input is captured by an operator or *gain function* g_u, which bounds how big the state can become under the influence of a signal of a given size. This plays the role of the transfer function operator in linear systems.

For instance, referring back to the motion of the ball in the basin, suppose the ball is in the equilibrium point **a**, and it is softly kicked. Intuitively, the ball will move away from the equilibrium but because it is a soft kick return and settle again at **a**. However if it had received a strong kick, sufficient to pass over the point **c**, the ball will never return. ISS can capture all of this by limiting the domain of where the Expression 6.13 is valid.

We may expect that an ISS concept based on the size of a signal will necessarily be conservative. The absolute size is not the only crucial information about a signal. On the other hand, because size is such a practical measure, ISS is very practical, in particular when considering the interconnection of systems. Roughly speaking, from this point of view, each system is replaced by a gain function, and the topology of the interconnection dictates how signal sizes are transformed. This allows us to readily verify if a system constructed from subsystems that are input-to-state stable keeps this property, and moreover it allows us to quantify how large the overall system's state can become a function of all the external inputs influencing the system. We explore this in the following sections.

Input-to-State Stability

A system, described by its state space model, either (5.18) or (5.17) is said to be ISS provided that for any initial condition x_0 and any bounded input function u (with bounded size $\|u\|$) the size of the state x can be bounded as

$$\|x(t)\| \leq g_x(\|x_0\|, t) + g_u(\|u\|), \quad \forall t > t_0 \tag{6.14}$$

The function g_x captures the transient effect of the initial conditions and it increases with the initial state and decrease to zero with time.

The function g_u captures the effect of the input size on the size of the state, it is called the input-to-state gain function. It must be non-decreasing.

ISS expresses a very strong, hence also very practical, and very desirable notion of stability. In particular, observe that for zero initial condition and zero input, both terms on the right side of the Inequality 6.14 are zero, which implies that the state is identically zero. In other words, $x(t) = 0$ is an equilibrium. Moreover, it is the only equilibrium.

There are ways of estimating the functions g_x and g_u for classes of non-linear systems based on ideas of Lyapunov stability. In fact ISS and Lyapunov stability are closely related.

For linear in the state and the input(s) systems[11], the notions of input-to-state stability and equilibrium stability coincide. Hence, as already mentioned, for linear system dynamics it is indeed quite legitimate to talk about a stable or unstable system. For nonlinear systems this is in general not at all the case, as we realized with the tank example as well as the ball in the basin.

Input-to-state stability is useful in that it allows us to infer stability properties for interconnected systems from knowledge of the ISS properties of the systems that are being interconnected. It is particularly suitable when the cause-and-effect direction is obvious, as the underlying implicit assumption in ISS is precisely that the information flow is from the input to the state, the former affects the future of the latter. When the information flow is clear, there are essentially but two main mechanisms for building large systems from subsystems, either by cascading two subsystems or by forming feedback loops of two subsystems.

Bounded-Input-Bounded-Output Stability

A particular case of ISS, precisely when the output coincides with the state, is the notion of Bounded-Input-Bounded-Output stability, BIBO for short. This concept expresses the idea that bounded inputs lead to bounded outputs.

6.6.3 Input-to-State Stability and Cascades

Consider a cascade of two systems. The idea is illustrated in Fig. 1.6. In a cascade of two systems the state, or an output of the first system ($y_1 = C_1 x_1$) serves as input to the second system.

If the first system S_1 (see bottom of Fig. 1.6) is input-to-state stable from input u to state (output) y_1 with input-to-state gain function G_1 and the second system (S_2) is also input-to-state stable from input (u_2) to state (output) y_2 with input-to-state gain function G_2, then the overall cascade with $u_2 = y_1$ is also input-to-state stable from input (u_1) to the state of the cascade which is (y_1, y_2). Moreover it is possible to estimate an input-to-state gain function. The gain function of the cascade is closely related to the composition of the gain functions of the subsystems, i.e., dealing with operators

$$G_{cascade} = G_1 G_2$$

Clearly, if both tanks separately are ISS, then the cascade will also be ISS.

---

[11] This means $f(x, u, t) = A(t)x + B(t)u$ in Eq. 5.18 or 5.17.

Interesting examples of cascades in system models have been encountered in Sect. 3.2 and 3.3, where we discussed the production of ceramic tiles and gravity fed irrigation systems respectively. In manufacturing processing, typically the output of each stage feeds into the next stage, creating naturally cascaded subsystems in the description of the overall system. Very similarly, in the irrigation channels, the immediate upstream water level will affect the water level in the downstream pool again producing a natural cascade of subsystems (pools) that describe the entire system (a channel). If each subsystem in isolation is input-to-state stable then we can conclude that the whole system is also input-to-state stable. Moreover we can quantify what effect a disturbance occurring in a subsystem will have on all downstream subsystems, and the system as a whole. In this sense input-to-state stability is a powerful concept. For example it is sufficient to understand input-to-state stability for a single subsystem in a cascade of identical or almost identical subsystems in order to deduce relevant information about the entire cascade.

> **Stability of Cascade Systems**
>
> To determine the stability of a cascade of systems we only need to analyze the stability of each subsystem: the cascade is stable if all the subsystems are. If any of the subsystems is unstable, the cascade is also unstable.

This is quite clear in the case of linear systems. The transfer function of the cascade is just the product of the transfer function of the subsystems, as can be easily derived by using the block-diagrams algebra (See 5.6.2). Thus, say for two systems in series, represented by

$$G_1(z) = \frac{N_1(z)}{D_1(z)} \quad \text{and} \quad G_2(z) = \frac{N_2(z)}{D_2(z)}$$

$G(z) = G_1(z)G_2(z)$ represents the cascade. Clearly the denominator of the cascade is the product $D_1(z)D_2(z)$, and the collection of its roots is the union of the set of roots from the subsystem transfer functions. The cascade is stable only if both systems are stable.

6.7 Beyond Equilibria

So far our discussion has concentrated on the stability of equilibria. In many engineering applications this suffices, but there are also many systems where the real object of interest is not an equilibrium but rather a time varying object. For example, in the antennae tracking problem considered in Sect. 3.4, the issue is to drive the antennae such as to track a star in the sky, whose position can be represented as a quadratic function of time. In other instances a system has to behave in a periodic fashion (like a clock, a wheel, a hard disk drive), or it naturally behaves in a periodic or almost periodic manner (like our solar system, the tide, our heart beat). In fact periodic and almost periodic phenomena are prevalent in nature and engineering systems alike.

The main tool that is used to analyze a time varying object is actually to attempt to map the object of interest to an equilibrium, perhaps for a different system. This equilibrium may be analyzed using the above techniques, in particular the input-to-state stability ideas, especially when the transformation into an equilibrium involve approximations. Finally our understanding is transformed back onto the original description for interpretation.

6.7.1 Limit Cycles and Chaos

Some systems may present different equilibrium points or even more complicated stationary behavior. Let us analyze a model for a population of rabbits (without preda tors). At a given moment, the number of rabbits is expressed by $x(k)$. We assume that the rate of birth is related to the number of animals and the rate of death, due to a lack of food (and old age), is proportional to the square of the number of rabbits. That is

$$x(k+1) = r.x(k) - d.x^2(k)$$

If we defined a new variable z, such that $z(k) = (d/r)x(k)$, the new equation[12] only depends on the parameter r:

$$z(k+1) = r.z(k)[1 - z(k)] \tag{6.15}$$

There are two possible equilibrium points, at $z_{e1} = 0$ and $z_{e2} = 1 - 1/r$ (which is only a realistic equilbrium for $r > 1$). Varying the parameter r, in this normalized model, leads from an overdamped stable equilibrium to oscillatory and even chaotic behavior.

For $0 \leq r \leq 1$ the evolution is overdamped and rabbit population goes to extinction, i.e. a final value of $z_e = 0$ is reached. For $1 \leq r \leq 3$ the evolution is underdamped and the rabbits do not become extinct (unless there were none to start with). The population reaches the final value of $z_e = 1 - 1/r$. At $r = 1$ there is a change in the local stability properties of the zero population. It is a bifurcation parameter.

For $r = 3$ there is another bifurcation, the so-called flip bifurcation. The system does not reach any equilibrium but instead it remains oscillating with period 2. For $r = 3.0$ and any initial normalized population $z(0) < 1$, the final population jumps between 0.6179 and 0.7131, as shown in Fig. 6.9a. Almost all populations will reach a *limit cycle* (apart from starting at the equilibria z_{e1} and z_{e2} which are now both unstable).

If we further increase the value of r, new oscillations appear and for $r \geq 3.57$ the behavior becomes *chaotic*. In the chaotic regime, the population evolution is very sensitive to the initial conditions, as can be seen in the graph depicted in Fig. 6.9b, where the coefficient $r = 3.8$ and the initial conditions are $z_1(0) = 0.5$ for the evolution marked with ○, and $z_2(0) = 0.5001$ marked with with ·. We realize that after 25 periods the evolution is completely different and the difference (marked by crosses) starts to be more and more significant.

---

[12] This equation is called the *logistic equation* and it shows very interesting properties related to bifurcations and sensitivity.

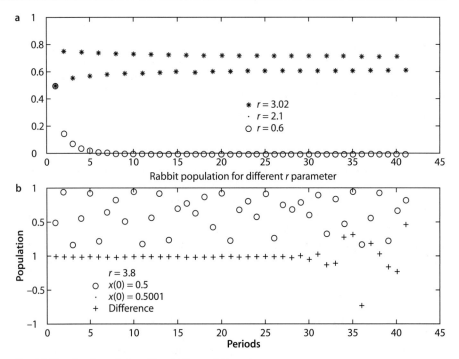

Fig. 6.9. The chaotic evolution of a rabbit population

This chaotic behavior could be interpreted as stochastic, but it is not. A chaotic behavior is totally deterministic, as the future state value can be exactly computed if the model and the current state are known. The main characteristic of a chaotic behavior is that the system evolution is extremely sensitive to small perturbations in the state value. From this point of view, a stochastic interpretation of this deterministic system may well be useful (there is sense in this madness).

Few systems exhibit such extreme *sensitivity*. In general, if the system is stable, small changes in either the system parameters, the input signals or new disturbances produce small changes in the system behavior. Qualitative changes, as marked by bifurcations, only appear in rare circumstances.

Because bifurcations are rare, understanding what can be and cannot be is important. Normal experimentation with a system may never show any trace of looming bifurcation points. Careful experiments must be planned to exhibit the phenomena of interest.

6.8 Sensitivity

In the example of the rabbit colony, the sensitivity of the behavior was with respect to an internal parameter (related to birth and death rates). To introduce the setting for the notions of sensitivity and robustness, we start with some additional simple examples.

Static Measurement/Sensor Inaccuracies

Let us consider a simple tachometer, as presented in Fig. 7.9. In principle the device is characterized by a static model such as $V = f(\omega)$, where ω is the rotational speed of the motor and V is the output voltage. If the tachometer axis slips relative to the axis whose rotational speed we are interested in, an error results:

$$V + \Delta V = f(\omega + \Delta \omega)$$

We call $S_\omega = (\Delta V)/(\Delta \omega)$ the input sensitivity function of the tachometer. In general, it is a function of the input. If the tachometer function is linear, the sensitivity function is constant and equal to the tachometer gain. This idea is represented in Fig. 6.10a.

Possibly, the actual measurement function is different from the one we have considered to be the model. That is, the measurement is $V = F(\omega) \neq f(\omega)$. This would be the result of an incomplete calibration of the instrument, for example. In this case we do not know the true function F, but we may know something about the difference, because we have information from the calibration of the instrument. Let the difference be

$$W(\omega) = F(\omega) - f(\omega)$$

Denote $\hat{V}$ as the expected value of the voltage, we can use a block diagram as shown in Fig. 6.10b, to represent the tachometer. The function $W(\omega)$ represents the uncertainty in the knowledge of the model, here represented as an *additive uncertainty*.

The notion of how close two functions are is not a simple one. One (conservative) measure for the difference between two functions could be identify it with the maximum possible deviation over all possible measurements.

Another model for measurement uncertainty is

$$V = F(f(\omega)); \quad V = f(F(\omega)) \tag{6.16}$$

Here F represents a function, presumably close to unity that captures the uncertainty. On the left the uncertainty acts on the output of the *ideal* device, on the right it acts on the object to be measured directly, i.e. the input of the instrument. The latter is called an input uncertainty, the former an output uncertainty. How much F deviates from unity indicates how little we know about the actual measurement instrument f. In a linear system context, this uncertainty is described as a multiplicative uncertainty. In either the linear or nonlinear case, it is represented as a cascade of systems, only in the nonlinear case the order of the sequence matters.

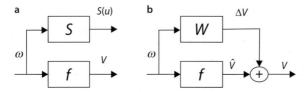

Fig. 6.10. Sensitivity of a static measurement device

In any measurement device there are many different sources that lead to errors. There may be systematic errors, or random errors due, for instance, to quantization. In all control applications, it is important to know about all the sources of uncertainty. More preferable is to have some information as to the possible size of these sources of uncertainty so that acceptable performance expectations can be quantified.

Unmodeled Dynamics

Let us consider a compact disk (CD) player. The laser beam is attached to an arm moved by a very fast motor (a voice coil motor or VCM). A simple model of the motor may be a double integrator, from force to position. In doing so we assume that the arm is rigid, that the disk is rotating at a fixed and known speed and that there are no external disturbances. The real system however has mechanical resonances as it is not perfectly rigid (fast movement, means light, means not rigid). Moreover the arm is impacted by external vibrations and the relative position of beam and track on the rotating CD is subject to the eccentricity in the disk, which will vary in magnitude from disk to disk. All these are disturbances away from the ideal model.

The way these disturbances affect the behavior of the CD player are quite different, mainly because their range of frequencies, spectral content, is very different. The eccentricity of the CD rotating has a perfectly well-defined frequency and can be tracked and compensated for. The mechanical resonances are typically in the very high frequency range and hence are only relevant in fast transient when moving the pointer from one track to another. Typically they are avoided by ensuring that the transients are not exciting the resonances. The external mechanical vibrations are much lower in frequency, related to the sound being produced out of the speakers for example, and the controlled system has to be insensitive to these disturbances.

This example is very much like the tracking antennae in Sect. 3.4.

If we go back to the tachometer, another kind of disturbances may be considered. The brush sweeping effect in the rotating machine will produce a high frequency noise. This noise, if not filtered, will be fed back to the process and will disturb the operation of the system to which the tachometer is attached. This issue will be considered in Chap. 7.

6.8.1 Robustness

Typically the term sensitivity is reserved to describe the effect of small variations in parameters in the model on a particular signal in the system, or the effect on such signals caused by small variations in external signals and so on. Robustness refers to a similar cause-effect relationship, but considers large variations belonging to some predetermined set of allowable perturbations. Nevertheless, both terms are often used interchangeably. In a colloquial way, robustness refers to a lack of sensitivity. Thus, a system signal, or system property may be robust with respect to variations in the input signal, meaning that the allowed perturbations away from the nominal input is not going to cause a significant departure in the response or property of the system. Or, a system is robust with respect to a system variation, i.e. the actual system is different from the model, but this difference has little impact in that the behavior of the system is pretty much like the behavior of the model. In each case robustness refers to the insensitivity of a particular property with respect to a perturbation that belongs to a predetermined set of perturbations to be considered.

Robustness is not a universally desirable property. A system is designed for a purpose, and responsiveness may be exactly what is required. Robustness must only hold in the sense that the desired behavior is insensitive to changes in the rest of the system, or signals impacting on the system.

Robustness of stability reflects to what extent system parameters are allowed to vary without affecting the desired stability. This is so-called *robust stability*.

Stronger than robust stability, because stability is really a must have property in a controlled system, is *robust performance*. In this case, the requirement is to keep the controlled system performing within acceptable limits, despite a range of variations in the system or external signals impacting on the system. The aim is to have only a minor change in performance, despite a large perturbation somewhere in the system.

By way of example, a hard disk drive has a guaranteed seek time, the maximum time it takes to go from one track to another. This performance specification has to be met by each hard disk drive mechanism. The control algorithm has to ensure this, despite the fact that the mechanical resonant frequencies may vary significantly between different units (even produced on the same manufacturing line). The seek time is robust with respect to resonant frequency.

6.8.2 Sensitivity Computation

Typically the following uncertainties or perturbations must be considered in system design:

- *External disturbances*, external signals entering the system, some measured (like the variable composition of the feed-material in the ceramic tile factory), some unmeasured (like the noise in the music recording, or the wind load on the antennae).
- *System variations in parameters*, often referred to as *structural uncertainty* or *parametric uncertainty* (like the resonance frequency in the hard disk drive).
- *System model errors*, often referred to as *unstructured uncertainty*. Due to the fact that the model does not capture all of the system dynamics. For example in the antennae servo design, the resonances are not part of the control model.

With respect to a single loop control system, as depicted in Fig. 6.11, the design engineer will verify the sensitivity or robustness of all the signals in the loop with respect to the above uncertainties. In Fig. 6.11 d_e represents an external disturbance, becoming d in the system, and r_f is a filtered reference.

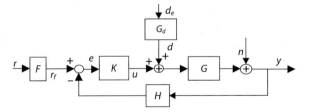

Fig. 6.11. A typical closed-loop control system

6.8.3 General Approach

When designing for robustness, the operational boundaries of a system are first defined. Typically no bifurcations are allowed within the normal operating envelope of the system. Bifurcations will be considered as part of the analysis of the behavior under fault conditions. Within the operational envelope of the system it is safe to assume that all cause-effect considerations, robustness and sensitivity can be captured through operators whose properties vary smoothly with parameters, signals and so on.

From the block diagram in Fig. 6.11, assuming for simplicity $F = 1$ and $G_d = 0$, and an ideal, unitary gain sensor, $H = 1$, the operators of main interest are:

$$T = S_{yr} = \frac{y}{r} = \frac{GK}{1+GK}; \quad S_{yd} = \frac{y}{d} = \frac{G}{1+GK}$$
$$S_{ur} = \frac{u}{r} = \frac{K}{1+GK}; \quad S = S_{yn} = \frac{y}{n} = \frac{1}{1+GK} \quad (6.17)$$

S_{yr} represents the response of the output y to the reference signal r, and so on. As can be observed, all these operators have the same denominator. We could think that just making this denominator somehow large enough all the sensitivities would be small, regardless of which signals we consider.

There are constraints though. In particular for the so-called *system sensitivity*, $S = S_{yn}$, indeed it is trivial to observe that

$$S_{yn} + S_{yr} = 1 \quad (6.18)$$

This equation tells us that system sensitivity $S = S_{yn}$ and the so-called *complementary sensitivity* $T = S_{yr}$, cannot both be small at once, as their sum adds up to the unity operator. We have to accept some sensitivity in a control loop. This is a fundamental constraint[13].

All sensitivity functions play an important role in control design.

6.8.4 Sensitivity with Respect to System Dynamics Variations

The Operators 6.17 describe the response of the system variables $y, u, \ldots$ to the external inputs $r, n, \ldots$. They define the performances of the system. In design, it is precisely these operators that are to be designed. It is therefore important to analyze what happens if the operator G (describing the system under control) changes, in this way we explore what happens to our design when the real system differs from the model used in design.

In particular, let us see the influence of changing G on the reference signal to output operator, the complementary sensitivity, T. Assume $H = 1$, for simplicity, and take the *derivative* of T with respect to G in Expression 6.17[14], it is easy to see that

---

[13] It should be noticed that the fundamental limit imposed by the equality $S + T = 1$ allows both S and T to be larger than one. Think of both as a complex number (that varies with frequency), or one positive and large and the other negative and large.

$$\frac{dT}{T} = S\frac{dG}{G} \tag{6.19}$$

That is, the ratio of the relative change in the closed-loop operator dT/T due to a relative change in the open-loop operator dG/G is precisely the sensitivity function S, which is also the response from the noise to the output. This underscores the importance of the sensitivity function.

6.8.5 Sensitivity Measurements

We expressed sensitivity using operators, allowing us to see how the sensitivity depends on particular system parameters. It is also apparent that there is an inherent trade-off due to the fundamental constraints, and sensitivities cannot be arbitrarily specified.

Information such as the maximum gain of these sensitivity operators is important information about the system behavior:

$$M_s = \max_\omega |S(j\omega)|; \quad M_T = \max_\omega |T(j\omega)|; \quad M_{yd} = \max_\omega |S_{yd}(j\omega)| \tag{6.20}$$

and are useful to characterize system sensitivity.

There are positive and negative forms of sensitivity. For instance, high sensitivity is important in measurement, to detect special signals, like the frequency selection in a radio. On the other hand high sensitivity may be dangerous, as in a hyper allergic reaction. Similarly high sensitivity to resonance or noise could be disastrous to the radio telescope or hard disk drive mechanism.

> **Summary of Stability and Robustness**
>
> The general concept of stability can be expressed as:
> *A system is stable if its behavior (which includes all initial conditions) has the property that for all bounded inputs, all conceivable signals are bounded.*
> We discussed different forms of stability:
>
> - *Local stability* of an equilibrium (orbit) in an autonomous system requires all system responses starting in initial conditions close to the equilibrium (orbit) to remain close to the equilibrium (orbit);
> - *Asymptotic stability* of an equilibrium, if the responses are both stable and asymptotically approach the equilibrium;
> - *Global stability* of an equilibrium in an autonomous system, if the above holds true for any initial condition;
> - *Robust stability*, if the stability property is maintained under changes in the system parameters;
> - A system response is *sensitive* to some parameters, input signals or operators if the system response varies (significantly) with changes in these parameters, input signals, operators.
>
> *For linear systems, stability is an intrinsic property of the system.*

---

[14] $\dfrac{dT}{dG} = \dfrac{K(1+GK) - KGK}{(1+GK)^2} = \dfrac{K}{(1+GK)^2} = \dfrac{T}{G} S$.

6.9 Comments and Further Reading

Stability is a well-studied topic, originally motivated by mechanics and such lofty questions as "Is our solar system stable?". The early work by Lyapunov laid the foundations for a more general study of stability. His treatise is still an excellent introduction to the topic (Lyapunov 1992). Stability is dealt with in detail in nonlinear systems texts such as Khalil (2002) and Willems (1970), which also provide a modern introduction to Luyapunov's first and second method. The celebrated Hartman-Grobman theorem, extending Lyapunov's first method is described in Guckenheimer and Holmes (1986).

The notion of input-to-state stability, and its connections to Lyapunov stability is very important in systems engineering as it allows us to deal with systems with inputs, and hence interconnection of systems. The work by Sontag is key in this area, see for example Sontag (1998).

Bifurcation theory is a branch of dynamical systems theory. It seeks to bring order in the domain of closed system behavior. Typical books include Wiggins (2003), Guckenheimer and Holmes (1986) and with a greater emphasis on computational ideas Kuznetsov (2004). One-dimensional dynamics, like the logistic equation, are well understood. The generic behavior of such dynamics as well as its robustness are comprehensively treated in de Melo and van Strien (1991). A dynamical systems approach to control systems is not for the faint hearted. Results in this direction are summarized in Colonius and Kliemann (2000).

Robustness and sensitivity are important notions in all engineered and general systems. These notions require extensive computational resources to be fully explored. There are fundamental constraints imposed by the system interconnection structure, and hence the many iterations in designing new systems. In the context of linear systems there is a well-established theoretical framework to deal with robustness and sensitivity issues (Green and Limebeer 1995; Boyd and Barratt 1991), even for large scale systems.

Chapter 7 Feedback

7.1 Introduction and Motivation
7.2 Internal Feedback
7.3 Feedback and Model Uncertainties
7.4 System Stabilization and Regulation
7.5 Disturbance Rejection
7.6 Two-Degree-of-Freedom Control
7.7 Feedback Design
7.8 Discussion
7.9 Comments and Further Reading

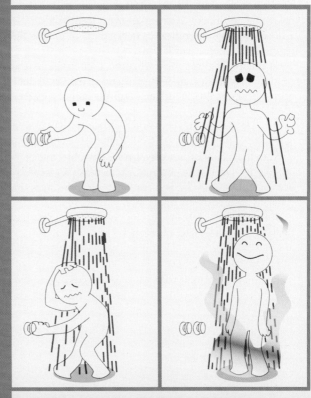

Chapter 7

Feedback

> *We need feedback*
> *to know how well we have done*
> *and to improve our future behavior.*
> *But should it be positive or negative feedback?*

7.1 Introduction and Motivation

The behavior of two or more systems connected in series (or cascade) is relatively easy to predict. The first system input is a free signal. Its output is the input to the next system, which in turn responds to this input, and so on, till we reach the end of the cascade. The behavior of a cascade is well-defined, as long as each sub-system in the cascade is well-defined[1]. At each stage the signal is affected by the system in that stage. Despite its simplicity, we know that a cascade does not always deliver as expected, as anyone who has ever played the game of *Chinese whispers*[2] knows all too well.

When systems are interconnected in a loop things are different. Even with two systems in a loop, where do we start? When do we stop chasing our own tail? Predicting what behavior is going to exist is no longer a matter of simply understanding the sub-systems in their own right. The interconnection is essential in defining the behavior. Moreover, where typically cascade connections are always well-defined, feedback connections are not. A well-known example of a feedback loop that is not well-posed is the acoustic feedback that occurs with having a microphone listening to the output of a speaker. Depending on the amplifier in the loop, the result can be most unpleasant. The shower example, illustrated at the start of this chapter, is another example of how feedback has to be treated with care.

Just consider two unitary systems interconnected in a positive feedback loop with external additive input $r = 1$. Doing some simple calculation you will come to the absurdity that $1 = 0$!!!, indicating that the feedback system is not well-posed. A little more generally, as discussed in Sect. 5.6.2, the basic operator in a simple feedback loop consisting of two systems G_1 and G_2, is given by Eq. 5.13, rewritten here

$$G = G_1(1 + G_1 G_2)^{-1} \qquad (7.1)$$

The issue of well-posedness is concerned with the existence of $(1 + G_1 G_2)^{-1}$. In our previous example this expression is $(1 + (-1)(1))^{-1}$ for which there is no valid interpretation.

---

[1] Tacitly, we are assuming that the outputs of the previous system in the cascade are acceptable inputs to the next system, but presumably nobody would build or even conceive a cascade unless this was the case.
[2] Whispering a small piece of information into a neighbor's ear, who then transmits the message to the next in line and so on. At the end of the line, the received message is compared with the original. Typically the final message bears little resemblance to the initial message!

In our discussions we are going to make a leap of faith, assuming that engineers and nature know how to build systems. We are going to simply assume that the feedback loops are well-posed.

Much of the power of feedback loops has to do with the $^{-1}$ (denoting the inverse) in the above expression. A feedback loop creates some kind of an inverse or a division of an operator. Just like we cannot divide by zero (the previous example told us as much), inverses of operators are tricky, but being able to divide or take an inverse by simply interconnecting some elements is pretty neat, and very powerful.

As observed before, feedback is based on the observation of an output to modify the input to the same system. For this reason, systems incorporating feedback are also known as *reactive* systems. From a control point of view, this is a drawback because the control system in order to react must first detect the output, or error condition, before it can take action. On the other hand, very little knowledge about the system behavior may be required to react appropriately. Moreover, feedback reacts to changes in the loop, sometimes this is referred to as an adaptive response. In contrast, in open-loop (feedforward or cascade) for the final output to do what we want it to do, the models of the systems in the cascade must be precisely known, so as to select the right input signal. Roughly speaking the desired input into the cascade must equal the output processed by the inverse of the cascade, not an easy task in general. Moreover, the input to the cascade cannot react to changes in any element of the cascade.

Feedback plays a key role in systems dynamics. It is ubiquitous in both the engineered and the natural world. In this chapter we consider some of the powerful properties feedback offers. We start with a number of examples to illustrate the pervasiveness of feedback. Next we explore the power of the feedback a little more.

7.2 Internal Feedback

In Chap. 5, we learned that natural and engineered systems alike are mostly composed of rather elementary systems such as gains, integrators (accumulators) and (transport) delays, amongst many. It is precisely the existence of feedback that bestows richness on the dynamic behavior of interconnected systems. Let us revisit some of the feedback loops we already encountered.

Filling a tank. In Chap. 2, a kitchen sink (Fig. 2.7) is illustrated. The water level is stabilized because there is feedback: a constant inlet flow increases the amount of stored water until it reaches a level that produces an outlet flow equal to the input flow. We often identify this situation with a negative feedback loop: the higher the water level, the greater the outflow, that reduces this water level. See also Fig. 7.1.

Exothermal reaction. Some chemical reactions become more active as the environmental temperature increases. Some of these chemical reactions are also exothermic, that is, they release heat. Such a reaction inside a thermally well-insulated vessel will be explosive as the temperature will increase as a consequence of the reaction, which in turns accelerates the reaction, which in turn creates more heat, and so on. In this case, a positive feedback determines the dynamic evolution of the system, which in this case will lead to an explosion of some kind as to destroy the vessel, and terminate the positive feedback loop.

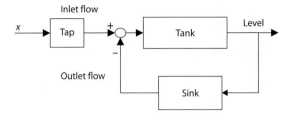

Fig. 7.1. Simple feedback loop

Electric circuit. Let us consider a simple loop circuit with a resistor, a capacitor and a battery. The voltage across the resistance is the difference between the battery voltage and that of the capacitor. As soon as a current flows through the circuit, the capacitor accumulates the extra charge and its voltage increases. Once the capacitor voltage equals the battery voltage the current ceases and the voltage stabilizes.

Motor. A voltage applied to an electric motor driving a fan generates a torque on its axis. This torque will accelerate the shaft, increasing the angular speed. A counter acting, reactive, friction torque appears as well as a mechanical load (from the wind flowing through the fan) that increases with speed. An equilibrium will be reached once the motor torque equals the mechanical counter torque.

Ecology. Remember the wolf and rabbits example in Sect. 1.4.4. On the one hand there is a positive feedback: exponential growth of the rabbit population without the predators, and on the other hand there is a negative feedback as the rabbits are eaten by the wolves. The interesting thing here is that these populations do not settle to an equilibrium, but rather evolve oscillatory, as in a limit cycle.

> **Internal Feedback**
>
> The dynamic behavior of a system is determined by both the dynamic behavior of the subsystems (gains, accumulators, delays and so on) as well the interconnection of these.
> Feedback is intrinsic to most physical systems.
> Compared to a cascade or series connection, a feedback loop made of the same components will naturally display a richer behavior.
> Feedback is used to **shape** (emphasize, modify) dynamic behavior.

Using some previous examples we illustrate how feedback modifies behavior:

Thermostat. The temperature in a room is stable when heat lost balances against heat gained. A thermostat will switch a heater on when the room is too cold and when the room reaches the right temperature it switches the heater off. This rather trivial on/off, negative feedback, controller based on some measurement of the room temperature is able to regulate the room temperature, as long as the heater is sufficiently powerful to reach the set temperature for the given room dynamics as determined by its size, the outside temperature, and the insulation (windows open or not).

Motion control. In playing any ball sport, our eyes provide positional feedback and assist us in directing our motion to intercept the ball, and/or provide it with the desired motion. Similarly in driving a car we use preview information from the road ahead of us to steer the car, but we rely on our reflexes (fast control loop) to avoid the dog that runs across the street.

From these examples we realize that we do not need a lot of information, nor a precise model of the plant behavior, to implement feedback (control). There is even model-free control based on feedback (like the thermostat). In many *simple* systems, negative feedback usually has a stabilizing effect (as in the tank filling, or the rotating motor) whereas positive feedback tends to make a system unstable (as in the exothermic reaction or the rabbit's colony from Chap. 1, or indeed the acoustic feedback).

In general though, to predict the dynamic behavior of more complex systems in a loop or with multiple loops is not straightforward. Feedback may create instabilities, but it also enables performance and behavior that is nearly impossible to arrive at otherwise.

Also, positive feedback is not universally bad. For example, oscillators, like clocks, are important and in biology as in the engineered world, positive feedback is exploited to create oscillators. In social networks, positive feedback encourages and can engender desired behavior in those that receive it (we all love praise).

7.3 Feedback and Model Uncertainties

The operational amplifier. An operational amplifier is an electronic circuit with a very large gain A (say $A > 10^5$) between its input and output voltage. The normal output voltage is measured in Volts, say less than 10 V and hence the input voltage will be less than 0.1 mV (10 V divided by a gain of more than 10^5).

When building circuits with operational amplifiers, we can rely on the fact that the gain for any one amplifier is large, but we cannot rely on the precise value of this gain. Indeed the variation of the gain across operational amplifiers even from the same manufacturing batch is also very large. A gain variation of a factor of 10 or more is not uncommon. Also, over the life of the circuit, the gain will change with time and the temperature of the circuit. The redeeming feature is that the gain stays large no matter what (say somewhere between 10^4 and 10^7). It also follows that that under normal behavior, the input voltage to the operational amplifier will always be negligible (less than 1 mV).

If an operational amplifier is connected in a negative feedback configuration, that is, the output voltage is looped back to the terminal labeled with a $-$, as shown in Fig. 7.2a, the gain in the new circuit, between the output V_o and the input V_i is:

$$\frac{V_o}{V_i} = -\frac{R_o}{R_i} \frac{1}{1 + \frac{1}{A}\left(1 + \frac{R_o}{R_i}\right)} \tag{7.2}$$

This relationship is represented in the equivalent block diagram Fig. 7.2b. Because the amplifier's gain A is very large no matter what, the input-output relationship is well approximated by

7.3 · Feedback and Model Uncertainties

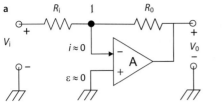
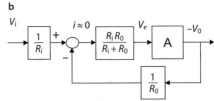

Fig. 7.2. An operational amplifier

$$\frac{V_o}{V_i} \approx -\frac{R_o}{R_i}$$

Observe that the actual gain of the amplifier does not matter, nor does the rest of the circuit, where V_o or V_i are connected into! (Amplifiers simplify circuit analysis and synthesis considerably!) The relationship between these two voltages is solely based on the resistors around the operational amplifier. When these resistors can be defined very accurately, so is the relationship between the input and output voltage. This is an important simplification, and all due to large negative feedback.

> **The Operational Amplifier**
>
> For those who know some circuit theory. With Ohm's law (the current through a resistor is proportional to the voltage drop across a resistor), and Kirchoff's laws, (that the sum of voltages in a loop is zero and the sum of currents at a junction is zero), you can readily derive the input-output relationship for the operational amplifier circuit. From Fig. 7.2a, at the summing junction 1, using Kirchoff's current law, knowing that the input current to the amplifier is zero we have
>
> $$\frac{V_i + \epsilon}{R_i} = \frac{-\epsilon - V_o}{R_o}$$
>
> The amplifier's equation is simply $V_o = A\epsilon$. Replacing ϵ by V_o/A in the previous equation, yields after some algebra, Eq. 7.2.

As shown in Fig. 7.2, whatever is in the direct path of the loop is not very important, as long as the gain is large enough. The whole system behaves as having a net gain of $-(R_o)/(R_i)$.

This result can be extended to any dynamic feedback system. Consider a block diagram from Fig. 6.11. Under the assumption that both the disturbance and the noise signals are zero, the following relationship holds:

$$y = \left(F\frac{GK}{1 + GKH}\right)r \qquad (7.3)$$

Thus, as far as the gain of the operator GK is much larger than unity, for the signals of interest, the input/output relationship is only determined by F and H (Eq. 7.4). Designing both elements accurately will deliver the desired response, independent of the plant or the feedback controller:

$$\|GK\| \gg 1 \Rightarrow y \approx FH^{-1}r \qquad (7.4)$$

It may look too good to be true. In general it is, although it is almost true. Indeed we are not able to make $GK \gg 1$ and also achieve stability or well-posedness for all possible reference signals r; but it is possible to have GK large where it matters[3], when r belongs to some class of signals, and it is then possible to get the above relationship holding where r matters.

The same concept can be useful when dealing with more complex system and operators (nonlinear, stochastic, multivariable and so on).

By way of illustration, suppose that r is a constant signal, $F = H = 1$ and there is neither disturbance nor noise. In this case having in the feedback controller K a pure integrator (which is retained in GK, which is the case if the plant does not differentiate its input) would guarantee that $y = r$ whenever steady state is reached. This is true assuming a well-posed feedback. Indeed in steady state y is a constant. From Fig. 6.11 the input to the feedback controller, K is then $r - y$ also a constant. Because the feedback controller contains an integrator, this constant must be zero. If it were not so, then this constant input would be integrated and this would yield an unbounded signal inside the loop, which contradicts the steady state assumption.

Compare this reasoning with the discussion of the toilet example. The cistern is the integrator in this case, and hence by the same argument, the cistern will always be filled to the right level (regardless of the other elements in the toilet).

For this reason, feedback controllers typically contain an integrator, as the integrator enables tracking of constant reference signals without an error. This is also called regulation. Having the plant output settled at a given reference level is an important task in control design, if not the most important one. As a consequence *integral action* is included in most industrial feedback controllers. In Sect. 9.3.3 this concept is elaborated on in the so-called PID controllers[4], which are literally everywhere in the process and manufacturing industry.

Clearly feedback can provide a good (although not always a perfect) solution for tracking, provided the loop gain GK is large and $FH^{-1} \approx 1$ where it matters for the signal to be tracked. These are weak requirements compared to what an open loop solution would have to do: $F \approx G^{-1}$. The feedback solution does not need a lot of information about the plant G in order to achieve good tracking (where r is important and G is small, K should be large so as to make GK large).

7.4 System Stabilization and Regulation

As already discussed, one of the main features of feedback is the potential to stabilize an unstable system. This also means that feedback can destabilize. In the example discussed in Sect. 6.4.3, α was the feedback parameter to be tuned and we realized that the system behavior could be adjusted (stable, unstable, oscillatory, damped). The system remains unstable if α is outside the interval $\{0, -2\}$. There is only a small window of opportunity for stability!

In a more physical setting, in a typical servo motor application, where the shaft must be positioned at a particular angle (think of the radio antennae problem with a

---

[3] For those who may see these relationships as representations in the frequency domain, GK must be large where the spectral content of r matters.
[4] PID, proportional, integral and derivative action.

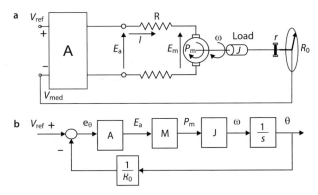

Fig. 7.3. A DC motor-based servo

fixed look direction) the variables of the system are as depicted in Fig. 7.3a. There is an electronic power amplifier/rectifier, a mechanical inertia, friction on the axis, as well as an integrator which links the motor speed to the shaft's angular position. The operator in the forward path from the drive voltage E_a to the position ϑ therefore contains an integrator. Any desired fixed angle position is clearly open-loop unstable, due to the integrator. If a voltage is applied, the motor will run, and the shaft will simply rotate. We can say that the system is unstable.

Let us introduce a negative feedback in such a way that the drive voltage is generated by, for instance, an amplifier as the one shown in Fig. 7.2a, with two inputs. The positive terminal has a voltage proportional to the desired position and the negative (feedback terminal) receives a voltage proportional to the current shaft's position. This is also shown in the block diagram Fig. 7.3b. As long as there is an error or a difference between these two inputs, a voltage will be generated by the amplifier and the motor will start to run, hopefully in the right direction so as to reduce the error (negative feedback). When the error vanishes, no voltage will be generated and the motor stops.

There will be problems if the amplifier gain is too high, because of the motor's inertia. As long as the motor receives a drive voltage it generates torque and rotates. With a large gain, even if the position error is small, this voltage is large, and the motor will have too much inertia to immediately stop or reverse direction even when the shaft went past the reference position. The drive voltage switches polarity, and eventually the motor starts turning backwards. The whole cycle repeats itself, producing unpleasant oscillations (vibrations). Feedback needs care.

As may be observed, in order to implement feedback, extra system infrastructure is needed: the components required to build the feedback path. In this servo motor example, the sensor could be just a simple potentiometer, transforming the axis position into a voltage. In other applications, the lack of appropriate measurement devices may prevent feedback from being considered.

This servo motor example, though rather different in its physical incarnation has all the features of the toilet cistern's behavior. The main features are the integrator (the motor or the cistern) and the negative feedback loop with a gain element (amplifier or float). Together these lead to a constant steady state equal to the reference (reference voltage or float position).

7.4.1 ISS and Feedback Systems

The concept of input-to-state stability considers stability of systems with inputs. This enables its use in system interconnections, and in particular feedback.

When considering a general feedback configuration of ISS systems, as depicted in Fig. 7.4 the overall system will be also ISS provided the total gain in the loop is sufficiently small. This is the celebrated *small gain* stability result.

In this figure, the external inputs are u_1, u_2, and the system states (outputs) are y_1, y_2 respectively. The internal signals are e_1, e_2. The meaning of the signals varies, but typically

- u_1 are external disturbances from the environment acting on the system of interest;
- u_2 are signals expressing control references and perhaps measurement errors;
- y_1 the state (output) of the system under control;
- y_2 the state (output) of the compensator, or control subsystem, our design freedom;
- e_1 the control input into the system under control;
- e_2 the measured feedback, derived from the system under control.

Small Gain Theorem

An important stability result associated with Fig. 7.4 goes as follows.

Assume that the ISS gain from input e_1 to output e_2 through system S_1 is given by g_1. This gain calculation is performed independent of the feedback loop structure, i.e. the input e_1 is assumed to be completely unrestricted. In the simplest case, say for linear systems, the gain tells us that the size of e_2 will be less than g_1 (a positive scalar) times the size of e_1.

Similarly assume that the ISS gain from input e_2 to output e_1 through system S_2 is g_2. Again, this gain function is determined independent of the feedback loop, g_2 is simply a property of system S_2.

Consider now the gain of the cascade S_1 after S_2 of the systems in the feedback loop. The gain of this cascade is the composition of the gain functions g_1 and g_2 and in the simplest case this gain is the product of the individual gains $g = g_1 g_2$. This gain is also called the *loop-gain*.

Conclusion. Input-to-state stability of the feedback loop, that is from the external inputs (u_1, u_2) to the external outputs/states (y_1, y_2) and also to the signals in the loop (e_1, e_2) requires that the loop-gain g is strictly less than unity.

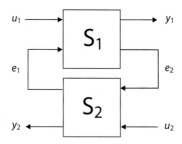

Fig. 7.4. A generic feedback loop

Gain Margin

The concepts behind the small gain theorem allow one to quantify a measure of stability for a system, or a degree of stability, denoted as the **gain margin**.

Consider in Fig. 7.4 a system G_1 (of interest) in a feedback loop, with a pure (scalar) gain system $G_2 = k$.

In general, as the gain k is increased, the feedback loop may become unstable. If the largest k such that the feedback loop is stable is larger than 1, then this is identified as the gain margin of system G_1. (1 plays a particular role, because traditionally it is assumed that the unity feedback loop is stable.)

This is perhaps counter intuitive in view of the small gain theorem, but systems may have an infinite gain margin.

More generally, the so-called conditionally stable systems are only stable for a given range of feedback gains k, say $k \in (k_{min}, k_{max})$. If zero does not belong to this interval (zero = no feedback) then such systems are open-loop unstable.

The small gain result is very useful, but also very conservative. Indeed it may well be that our estimates of the gain functions are not very tight, and these estimates may fail to meet the small-gain stability condition and yet the feedback loop may be stable and very well-behaved. Even if our gain estimates are tight, we should not expect that the small-gain condition captures all possible stable feedback loops. After all the gain functions only capture the effect of the size of a signal, and there is of course much more to a signal than just its size.

7.4.2 Linear Feedback Systems

In the case of linear systems, the question of stability and input-to-state stability can be settled using algebra for which there are efficient computational tools.

Let us reconsider Fig. 5.8 which is a special case of Fig. 7.4 for linear systems. It will become clear that feedback easily alters the stability properties and that the feedback loop can be stable and yet both systems in the loop unstable. Let the forward operator from u_1 to y_1 be $G_1(z) = N_1(z)/D_1(z)$, and the feedback operator from u_2 to y_2 be $G_2(z) = N_2(z)/D_2(z)$ (with negative feedback, as defined in Sect. 5.6.2). Using block diagram calculus, it is easy to see that the transfer function from u to $y = y_1$ is

$$G(z) = \frac{G_1(z)}{1 + G_1(z)G_2(z)} = \frac{N_1(z)D_2(z)}{N_1(z)N_2(z) + D_1(z)D_2(z)} \tag{7.5}$$

The roots of the denominator, $N_1(z)N_2(z) + D_1(z)D_2(z) = 0$ are clearly different from those of the components D_1 and D_2. That is, stability of the subsystems G_1 and G_2 does not imply stability of the feedback system. Neither does stability of the feedback system imply stability properties for the subsystems[5].

By way of example consider G_1 as a pure integrator in continuous time, $G_1(s) = 1/s$ with a feedback system which is a pure scalar gain $G_2 = g_2/1$. According to the above calculation, the closed loop stability is determined by $g_2 + s$, so that the feedback system is stable if $g_2 > 0$ (see also the discussion around Eq. 6.7). The open loop system G_1 is

---

[5] This will be a key point in designing feedback controlled systems, as discussed in the next chapters.

unstable. The integrator's gain margin is infinite. Applying a negative gain in G_2 has the effect of applying positive feedback, which indeed destabilizes the loop.

Similarly, in discrete time, with $G_1 = 1/(z-1)$, an integrator in discrete time and a feedback system $G_2 = g_2$ a pure gain. The feedback loop is stable provided the roots of $g_2 + (z-1) = 0$ are less than one in magnitude. The only root is $1 - g_2$, which is less than one in magnitude for $g_2 \in (0, 2)$. Again, the open loop system is not stable. The gain margin is finite, it equals 2. A negative gain as well as a gain larger than 2 will lead to instability in this case (remind the example in page 161).

Bode plots and Nyquist plots examine the behavior of the loop gain $G_1 G_2$ to identify the boundary of stability and this explains to some extent the popularity of the frequency domain methods, as before computers appeared, graphical methods provided powerful design methods. The key in this frequency domain approach is to ensure that at no time the loop gain $G_1 G_2$ can have a magnitude larger than one and a phase shift of 180 degrees.

Using the frequency domain ideas, it can be seen that the instability for one or two integrators described in Sect. 5.7 require positive feedback.

This was obvious in the single integrator loop, Fig. 5.14 where we explicitly used positive feedback to obtain an unbounded response (as explained just above again). Using frequency domain ideas, considering sinusoidal signals as input, it is clear that the integrator produces a sinusoidal output, of the same frequency, but with a phase shift equal to $-\pi/2$,[6] regardless of the frequency. It then follows that the loop gain can be arbitrarily large, and no instability will develop.

Positive feedback can however occur in a loop, even when we think we are applying negative feedback. As previously discussed, the elements in the loop will determine a loop gain and phase shift. Assume that for some frequency there is a phase shift of $\phi = -\pi$ radians or -180 degrees. That means a change in sign. If the gain at this frequency is higher than one, the effect is similar to a positive feedback and the closed-loop system is unstable. If the gain is exactly one than a resonance occurs.

This observation can explain loop instabilities and forms the basis of the celebrated Bode and/or Nyquist stability criteria.

In the two integrator case, Fig. 2.13 without friction, we have exactly -180 degrees phase shift. Each integrator produces -90 degrees phase shift, so the total phase shift is $\phi = -180$. The gain in the loop is $1/\omega^2$, which is one for $\omega = 1$ (in this way we have identified the resonance pulsation).

For the three integrator problem, any constant feedback will result in an unstable system. To illustrate the effect of the feedback, we use a small trick. First we apply some small negative feedback loop around each of the integrators, so that we have a nice and stable behavior in the forward path. This is illustrated in Fig. 7.5. The phase shift for each of these systems varies from 0 to 90 degrees. Hence in the forward path the total phase shift varies from 0 for low frequency to 270 for high frequency inputs. (The phase shift of a linear systems consisting of two linear systems in series is indeed the sum of the phase shifts of the linear systems.)

Assume unitary feedback around the three-integrator system. The smaller the feedback gain we use internally in the components, the larger the gain in the forward path. This implies that at 180 degrees phase shift the loop gain can be larger than one, pro-

---

[6] A phase shift of $\pi/2$ or 90 degrees, a cosine wave input produces a sine wave output.

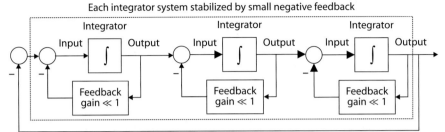

Fig. 7.5. Series connection of three feedback stabilized integrators, in a unity feedback loop; unstable for sufficiently small feedback gains

vided the introduced feedback gain is sufficiently small. Hence instability will result. The situation does not change when considering the limit, that is the case of zero feedback around the individual integrators, which is exactly the situation of three integrators in series in a negative feedback loop.

7.4.3 The Nyquist Stability Criterion

Proper design of feedback is a challenging problem in many real applications. One of the pioneers of feedback synthesis was Nyquist, who was faced with stabilizing telegraph telecommunication lines that required multiple signal amplifiers. He developed design rules, and the *Nyquist stability criterion* which settles the stability of single loop feedback systems for linear single input single output systems.

To capture the essence of this celebrated result, consider a unity negative feedback loop ($G_2 = 1$) with a plant in the forward path G_1 as in Fig. 5.8c, which is stable in itself. When is this loop stable? When is $(1 + G_1)^{-1}$ well-defined?

The problem can be appreciated as follows. Suppose that the input is a sinusoid. Because of linearity (and the fact that the plant is stable), the response is also sinusoidal with the same frequency, but with some phase shift and an amplitude change. In fact, all the signals in the loop will be sinusoidal in steady-state. If, for the applied frequency the open loop plant response is exactly 180 out of phase with the input with a gain of at least 1, then our negative feedback loop becomes a positive feedback loop. Because the gain is 1 or larger, chasing the signal around the loop indicates instability. If the gain was less than one, the loop results in some finite amplification, and all is well.

In general, the input of a system is not purely sinusoidal, but any signal can be decomposed in sinusoidal components (its power spectrum). Thus, if we analyze the frequency response of the plant (G_1), it will be easy to detect if for some frequency, the gain is larger than one and the phase shift being -180 degrees. If this is the case, the feedback system would be unstable. Graphically if we construct the (polar) plot of G_1 as a function of frequency, (an example in Fig. 7.6), in the complex plane the point $(-1, 0)$ must stay to the left, or remain outside the Nyquist contour[7].

---

[7] In the Nyquist criterion, a closed graph is plotted, the Nyquist contour, and the stability is determined by looking at the possible encircling of the $(-1, 0)$ point.

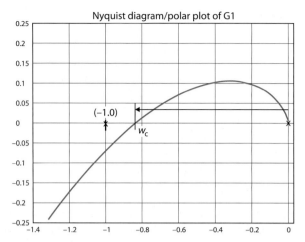

Fig. 7.6. Polar plot of a frequency response, from $\omega = 0$ (left) to $\omega \to \infty$ (origin). As the point $(-1, 0)$ is to the left of the curve, a unit feedback loop is well-defined

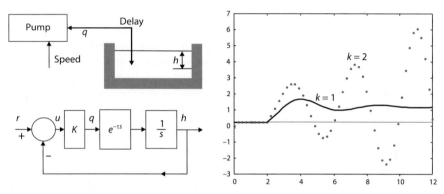

Fig. 7.7. Integrator with delay in a negative feedback loop

7.4.4 Integrator with Delay and Negative Feedback

Once the danger of positive feedback is appreciated, it is easy to see that systems with delay can cause havoc (like our shower!). A delay of τ seconds introduces a phase shift of value $\omega \tau$. Even a simple integrator with delay in a negative feedback loop can lead to instability when the gain is too large. This is illustrated in Fig. 7.7. Assume a simple tank, the net input flow q being controlled. The control signal is proportional, k, to the error in the tank level h with respect to a given reference r. It is not difficult to find when a gain is too large. The loop gain is k/ω. The total phase shift is $-(\omega + \pi/2)$. Therefore a phase shift of $-\pi$ occurs when $\omega = \pi/2$. The loop gain must be less than 1 at this frequency, so k is limited to be less than $\pi/2 \approx 1.5$. In the example in Fig. 7.7, a gain of 1 presents stable behavior, but a gain of 2 leads clearly to an unbounded response. In this case, the tank will either empty or overflow. Of course we will not see an unbounded signal, after all that would signify the end of the world as we know it.

7.5 Disturbance Rejection

Most systems are composed of many subsystems. This opens the possibility that in the connection between subsystems external, undesirable signals can enter. As the proper operation of the system depends on the information flow between systems, disturbances create problems.

Many different disturbances can be distinguished. In the case of signal transmissions they are in the main noise indicating some interference with or perhaps partial loss of signal. In other systems, as schematically represented in Fig. 6.11, the disturbances (signals d and n) may be more severe either corrupting a command signal or a measurement signal. Often it is either too difficult or too expensive to try to measure all possible disturbances. In this case, feedback can help to reduce the effect of these disturbances on the variables of interest. We say that feedback helps to make the system response *robust* with respect to the disturbances.

By way of example, consider an industrial boiler or steam generator (the central heating system in a building, or in a electrical power generating plant), as in Fig. 7.8, designed to provide steam (to heat, or to feed an engine).

In order to maximize the efficiency of the boiler it should work under design conditions of water level, temperature and pressure. If the boiler is going to run continuously with the same load, it may be enough to set some prescribed variables and let the system operate. Normally though the engine downstream or the building to be heated, does not need the same amount of steam all the time. The demand for steam is variable, and as a consequence the boiler should follow as to produce steam to meet demand. This has implications for feed water supply to the boiler and fuel supply to the burner. Steam demand is difficult to measure, and often somewhat unpredictable. Feedback comes to the rescue. Based on, for instance, the measurement of the boiler temperature, the fuel supply to the burner can be manipulated automatically. Feedback will react, if properly designed, in such a way that a decrement in the temperature will result in an increment of the fuel flow and vice versa. The influence of load variations will be greatly reduced and the temperature will remain close to the required value. Similarly the water level can be used to control the water supply, as shown in Fig. 7.8.

Again, by recalling the block diagram in Fig. 6.11, suppose that the disturbance d represents the steam demand, or load on the boiler. Without feedback, the variation in the

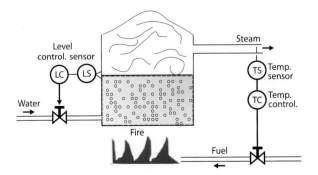

Fig. 7.8. Steam generator with some associated instrumentation

temperature (the output y) will be G times this disturbance. On the other hand, the relationship between the temperature and load variations when using the feedback will be

$$\frac{y}{d} = \frac{G}{1+GKH} \tag{7.6}$$

and again, if the loop-gain of the composed operator GKH is much larger than one for the signals we are interested in this relation reduces to simply $1/KH$. The controller K can then be chosen to achieve a required disturbance rejection, i.e. K is large where d is large. For example if d were constant, an integrator in K will reject the load disturbance completely i.e. it is not necessary to know the amount of steam required. The boiler will deliver it!

7.5.1 Noise Feedback

One unavoidable aspect of feedback is that the system input will be influenced through the measurement on which the feedback acts. This implies that whenever this measurement has an error the feedback will produce erroneous activity. Any measurement implies some deviation between what had to be measured and what is measured, as no sensor is perfectly accurate. Say that the measured signal is the desired signal plus noise.

This noise will therefore excite the system unavoidably. This could lead to undesirable behavior. For example if measurement noise could excite the resonances in the antennae structure (see Sect. 3.4), the radio telescope would be a useless instrument. In Fig. 7.9, the speed of a motor is measured by a tachometer providing a noisy measurement.

Hence we need to measure as well as possible or, failing to do this, to filter the measurement as to suppress the noise and yet retain the necessary feedback information.

Filters, observers or estimators are different forms of elements in the feedback path that extract from the noisy measurement as much information about the system/signals as possible. Filters can exploit models of the system dynamics as an advantage, so as to be able to distinguish a signal that could come from the dynamics from another one which is external to the dynamics. Sometimes, the information that is extracted is related not only to the signal that is measured but also to other signals internal to the

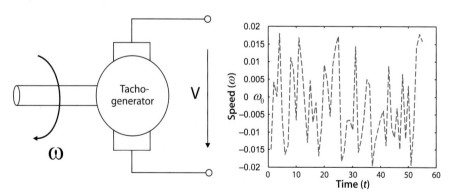

Fig. 7.9. A tachometer and the noisy signal it delivers

system, like the state components, as this may make it easier to compute what feedback is actually necessary. We will revisit this in the following chapters.

Looking at Fig. 6.11, the sensitivity functions previously defined (Eq. 6.17) are

$$T = S_{yr} = \frac{y}{r} = F\frac{GK}{1+HGK} \tag{7.7}$$

$$S = S_{yn} = \frac{y}{n} = \frac{1}{1+HGK}$$

Under the ideal conditions, ($HKG \gg 1$), these operators will be

$$T \approx \frac{F}{H} \; ; \quad S \approx 0 \tag{7.8}$$

Thus, whenever the feedback filter H has a low gain at a given range of frequencies in order to reduce the effect of noise, the system output/reference will necessarily have high gain or amplify the input signals in this range of frequencies. This identifies another clear trade-off between requirements with respect to noise rejection and reference tracking.

> **Control Performance Trade-Off**
>
> There exists a fundamental trade-off in any feedback control system:
>
> *a high tracking accuracy inevitably reduces the capacity to reject (similar) disturbances.*
>
> Because typically the signals we want to track (steps, ramps, sinusoids) are quite different (at least from a frequency content point of view) from those appearing as output disturbances that must be rejected, this trade-off can be negotiated by "strategically" allocating tracking accuracy in the right frequency band.
>
> Moreover, feedback control can be complemented with feed forward as in a two-degrees-of-freedom control strategy to improve overall performance (see also Sect. 8.6.1).

7.6 Two-Degrees-of-Freedom Control

If there are unwanted external signals, or disturbances acting on the plant, feedback can be designed to reduce their effect on the signals of interest. This happens without the need to measure the disturbance directly, and the feedback acts on the effect the disturbance has on the signal of interest (which is always measured, otherwise there would be no feedback). Well-tuned feedback reduces the sensitivity of the plant response to disturbances.

This property is very hard to achieve with feed-forward control, because to reduce the effect of a disturbance through the input to a cascade requires either precise measurement of the disturbance or precise (preview) knowledge. Also, in an open loop situation, if the systems in the cascade vary, this automatically affects the output. Using feedback it is possible to make some aspects of the response less dependent on these system variations, and thus maintain performance.

Consider again the typical control feedback system already discussed in the previous chapters (Fig. 6.11). The plant or system to be controlled is characterized by the operator G which is partially known, nonlinear, and perhaps time varying. The system is subject to an additive input disturbance d, which cannot be measured. For simplicity, let us assume $G_d = 1$. The plant output is affected by a (noise) disturbance n. The measurement device has a transfer operator H, which is used to suppress or filter out the effect of the disturbance n in the feedback loop. The feedback controller is K. A reference signal r is to be tracked by the output, i.e. we desire $y \approx r$.

From a design point of view, the questions revolve around the selection of K the feedback control, F the feed-forward control and to a lesser extent H, the measurement noise filter. The fact we have both K and F to work with, is captured as *two-degrees-of-freedom control*.

Feedback can be used to stabilize an unstable plant, attenuate disturbances, reject plant variations, and improve tracking performance. On the other hand, as already alluded to, too much feedback could destabilize the loop. Another unwanted effect is that feedback acts on what has been measured. So an incorrect measurement (as a consequence of the signal n in Fig. 6.11) will lead to unwanted control action.

Once feedback K (and H) has been selected mainly for responsiveness and disturbance rejection purposes, the tracking performance can be modified by means of F. As F is outside the loop, it does not influence the loop behavior too much, and does not affect stability at all. In this way two-degrees-of-freedom control can achieve both responsiveness and tracking performance. Because the feedback design makes the response less sensitive to plant variations, feedback is designed first, and then for the resulting compensated system the feedforward control is designed.

7.7 Feedback Design

Feedback may be used to get a controlled system that performs better than the open-loop system. Stability, tracking, disturbance rejection, are all aspects feedback may be used for. When one has the option of designing a new system that must have certain characteristics it pays to consider both the plant to be controlled and the feedback controller together to achieve the overall purpose. Of course, actually quite often, control arrives as an afterthought. The plant has been designed, and now some new requirements come forward, and feedback is added to an existing system. This has clear drawbacks.

Let us consider that you want to design a boat to sail in a race, fast and safe. First you could design a boat to be really safe, strong and able to master all winds and waves. This boat is presumably going to be rather heavy and hence sluggish. No amount of feedback used to optimally steer this boat, and trim the sails to the wind are going to make this boat behave like an America's Cup winner. More appropriately we would design a hull with a low keel, and little friction and large sails. The disadvantage is that wave action will be able to destabilize the boat, creating large oscillations, that would be virtually impossible to control by trimming the sails. Indeed this could not be guaranteed, as nobody can guarantee wind. Thus, creating a boat with an actuated keel, one that could be deployed and adjusted according to the wave action, and large sails that can be automatically trimmed would be the way to go. Not surprisingly the combined design of control system with plant to be controlled, leads to a superior solution.

Quite a similar situation may happen in designing the suspension of a car composed by dampers and springs. If they are passive elements, the stiffer they are the less displacement in the car body, but also the less comfort for the passenger. You may design them being very soft, but then large and sustained oscillations will result on a bumpy road. Feedback may provide the option to adjust the damping depending on the road conditions and the driving behavior: so-called active suspension.

In order to use feedback in systems' design as an advantage, the following conditions must be satisfied:

- well-defined goal(s) must be articulated;
- variables of interest must be measured, sensor subsystems must be provided;
- sensor information must be able to be related to goal attainment;
- the control algorithm must be able to decide how to take action, given information from the sensors and the goal(s), as well as the system model;
- feedback action can be applied as required to the system input, appropriately dimensioned actuators must be provided.

Open Loop Vs Closed Loop Control

Summarizing, open loop, as in a cascade or feed-forward control; closed-loop as in feedback control have the following characteristics:

- Open loop system controls cannot affect stability, those in closed loop can.
- Open loop systems are always well-posed, closed loop systems not necessarily.
- Open loop systems are not reactive, closed loop systems are.
- Open loop systems do not reject plant disturbances, closed loop systems can.
- Open loop systems are sensitive to plant variations, closed loop systems can suppress plant variations.
- Open loop control systems require accurate plant model and disturbance preview for performance, closed loop control systems do not rely on either an accurate plant model nor disturbance preview for performance.
- Open loop systems are not responsive to plant measurements, closed loop systems rely on accurate plant measurements.
- Combined feed-forward and feedback systems, two-degrees-of-freedom have all the advantages of either approach, and none of the disadvantages.

7.8 Discussion

Feedback is the most important feature in any controlled system.

Feedback is present in most natural as well as engineered systems. This follows because the basic building blocks in nature, as well as in the engineered world, are essentially very simple: accumulators (reservoirs, energy storage), gains, summers and delays. The behavior of these simple building blocks is nearly always trivial. The complex behavior so clearly observed in many systems, only follows through the interconnection of many simple subsystems using feedback loops, since without feedback, without creating loops, there is no complexity at all in the behavior.

Let us summarize the key ideas, and observations, together with some warnings:

- Feedback may be used to stabilize or destabilize a system.
- If the loop gain is less than unity, the closed loop is stable.
- Feedback may be used to improve robustness. To accomplish this high gain feedback is essential. If there is a high gain in the loop, the process model and or disturbances are less relevant and the global behavior mainly depends on the elements in the feedback path, which are specifically designed to achieve the control objective.
- High gain may result in components, like actuators, to saturate. High gain increases the risk for instability. There is a trade-off between robustness and stability.
- There is a trade-off between tracking and noise rejection.
- Two-degrees-of-freedom controllers help. The stabilization and disturbance rejection activities are tasks for the feedback loop, which is to be designed first. The tracking response is the task of the feed-forward controller, which can be designed next, based on the feedback stabilized loop.
- Feedback control reacts to errors detected in the system. Dealing with non-minimum phase systems (that is, systems showing an initial inverse response) or time delayed systems (that is, without any immediate response) require more sophisticated design. Simple high gain solutions will not work.

7.9   Comments and Further Reading

Feedback is a key concept in engineered systems and natural systems alike because of its ability to create interesting behavior from simple building blocks. As observed, feedback can significantly modify behavior. Feedback has a long history (Cruz and Kokotović 1972).

The importance of feedback in life as we know it, is well-established. See, for instance Hoagland and Dodson (1995), which gives a very gentle introduction. The importance of oscillations, and the role feedback plays to establish these in biology are well-documented in Goldbeter (1997).

A history of feedback control, with the emphasis on feedback starting with some interesting examples of clocks and float mechanisms is presented in Mayr (1970).

The input-output methodology is developed in Desoer and Vidyasagar (1975), and also in Mees (1991), which is heavily motivated by biology. Classical frequency domain ideas, building directly on the early work of Nyquist (1932) followed by Bode (1945) are developed in detail in Doyle et al. (1992) and a modern treatment using an algebraic and optimization-based approach is Green and Limebeer (1995). Despite the fact that a number of fundamental constraints have been established in feedback, we have only discussed superficially a few. No comprehensive or unifying treatment of feedback from this perspective exists in literature (even for linear systems).

A more mathematical treatment of feedback but still intended for a large audience, is Astrom and Murray (2008).

A treatise of feedback in a nonlinear systems and input-output setting is still under development. The role of input-to-state stability is developed in Sontag (1998) and Khalil (2002). Substantial advances have been made using ideas of passivity which we have not discussed. Passivity expands on the notion of system gain, and brings in the idea of phase, which plays such an important role in the linear system setting. A comprehensive treatment of this notion and how it applies to a very large class of nonlinear systems that can be described using the physical principles of electro-mechanical systems is Ortega et al. (1998). A more mathematical treatment is van der Schaft (2000).

The importance of feedback in electronic design is undisputed. A particular discussion of just this aspect of electronic design is Waldhauer (1982). The operational amplifier and its role in modern electronics is taught in all electrical engineering curricula.

The main control module used for feedback in the process industry is the so-called PID regulator, a good overview of the design questions surrounding PID controllers can be found in Astrom and Hagglund (2005).

Chapter 8: The Control Subsystem

8.1 Introduction and Motivation
8.2 Information Flow
8.3 Control Goals
8.4 Open-Loop
8.5 Closed-Loop
8.6 Other Control Structures
8.7 Distributed and Hierarchical Control
8.8 Integrated Process and Control Design
8.9 Comments and Further Reading

Chapter 8

The Control Subsystem

The illusion of freedom.
The feeling of being under control.
Who is controlling who?

8.1 Introduction and Motivation

In considering a natural or human made system that operates as expected, it is often difficult to differentiate between process and controller. Is the teacher in control of the class, or is the class in control of the teacher? Even simpler, when flushing the toilet there is the clear understanding that the action of starting the process, be it by a proximity sensor, or pressing a button, or pulling a chain achieves the desired effect: the toilet is cleaned, and ready for its next use. Nobody really cares about the internal workings of the toilet system. From this perspective, it is difficult to realize that a feedback loop is at work, let alone that certain references such as flushing time, and fill level have been set.

In some other cases, despite the fact that control and process are well-integrated it remains relatively easy to identify what is in control and what is controlled.

A steam engine in a locomotive is such an example. There is the combustion chamber where the fuel is burned, producing steam in the boiler, which through its pressure and expansion drives the crank and slider mechanism that turns the wheels. There is also a Watt's governor[1] as shown in Fig. 8.1. This device opens or closes the steam supply according to the actual wheel speed, too fast and the steam supply is throttled, too slow and the steam supply is opened up. It ensures that the locomotive runs at a nearly constant speed, as prescribed by the driver.

In this case it is very easy to distinguish two subsystems: the energy transformation chain from the chemical energy in the fuel to the kinetic energy of the train, and the governor

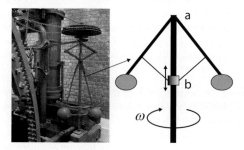

Fig. 8.1. Steam locomotive and Watt's governor

---

[1] The centrifugal force on the balls turning around a due to the speed of the train raises the point b and reduces the steam supply. James Watt, 1736–1819, Scottish engineer whose role in the industrial revolution cannot be overestimated. The unit of power is named after him.

regulating the speed. Steam engines would not run safely without a governor, neither would wind mills for that matter (it is from the latter that Watt adapted the idea to suit his engines).

The structure of this example is repeated in natural as well as engineered systems.

Control is now so pervasive that it is hard to consider any device that does not contain some form of control. Hard disk drives and compact disk players have track seeking and servo control. Cameras have image stabilization. Washing machines have sequencing, and often rule-based controllers that minimize energy and water usage as dish washers do. Microphones and hearing aids and recorders have automated gain control to avoid saturation and improve the fidelity of capturing sound. Cars are full of control devices, some have brake-by-wire, traction control, cruise control, climate control, engine control, lights and screen wipers that respond to the driving conditions, adaptive suspension and so on. Similarly aircrafts, trains, ships … Nowadays driven by a need for greater efficiency or more sustainability there is a push for pervasive networking, creating smart infrastructure (water, electricity and gas distribution systems as well as transport networks) and developing the largest networked control systems spanning entire continents.

So far we have considered the simple interconnection of systems as *open loop* or *closed loop*. This chapter goes somewhat beyond this, and is particularly concerned with the control subsystem, what it consists of, and what we may expect from it, and how it is interconnected in the overall system.

Historical Note

In its early incarnations, feedback was an art, an inherent part of making things work. The earliest examples of feedback are probably found in the first mechanized ancient water clocks (clypsydra) found in ancient Rome and Greece, as early as 300 BC. The water clocks, which essentially keep time through a constant flow of water, were made more precise due to a pressure regulator in the water supply. Another well-known example of feedback art, again related to time keeping, is the design and indeed evolution of the escape mechanism, essentially a speed regulator that brings the swinging motion of the pendulum over to the motion of the hands of a mechanical clock. All of these were implemented by trial and error, without much theory or prediction (which comes with analytic design) of how they would function.

The trial and error approach failed at the dawn of the industrial revolution. Indeed, the steady state oscillations or *limit cycles* that appeared in some applications of steam engine governors lead to the first systematic study of stability, with contributions by Maxwell[a], Routh[b] and Hurwitz[c]. Similar governors were in use well before the steam engine, to regulate the speed of wind mills all over England and the European mainland. They functioned well, but the interaction with the much faster working steam engine created dynamic instabilities not encountered before. It took about from the middle of the 19th century to the middle of the 20th century to transform the art of feedback into an engineering discipline. The introduction of digital computing enabled automation to proliferate, and now control engineering is a normal part of engineering curricula in most engineering disciplines all over the world.

---

[a] James Clerk Maxwell, 1831–1879, best known for the Maxwell equations describing the electromagnetic field. One of the greatest minds of all time, according to Einstein.

[b] Edward Routh, 1831–1907, English mathematician, best known for his work on mechanics, and an analysis of stability of linear systems.

[c] Adolf Hurwitz, 1859–1919, German mathematician which become famous for his work on algebraic curves.

The role of the control unit is intimately related to how it is interconnected with the object under control. Its position in the cascade or feedback loop conditions how the control acts in the overall system, and this in turn affects how it is designed. The actual nature of the control unit can be mechanical, electrical, electronic or digital, depending on the most appropriate technology and the application. The different elementary ways in which the control subsystem can be linked with the system are shown in Fig. 8.2.

> **Control System Interconnection**
>
> a *Open loop, control before the plant.* The control unit is to generate the appropriate command inputs for the plant.
> b *Open loop, control after the plant.* The control unit evaluates the output of the plant and decides about future inputs for use at a later stage. Typical in batch processes and so-called quality control, not for real time control.
> c *Feedback loop, feedback control.* The control unit generates the inputs for the plant, based on real time information about the response of the plant, and information external to the plant.

More complex structures can be conceived, but they all use the above as elementary building blocks.

These different options will be further explored in Chap. 10, where we discuss design aspects of the control system. The purpose of this chapter is to analyze the structure of the control subsystem itself. What is it for? What are the options? What can we expect from it? We already know the main concepts to be discussed in this chapter. They are the information flow, the control objectives and the control constraints, feed forward versus feedback, mixed control, integrating process and control.

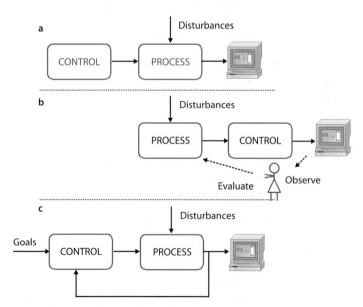

Fig. 8.2. Process and Control subsystem interaction

8.2 Information Flow

At a basic level, it is clear that monitoring and control both depend on information extracted from the process (perhaps through sensors) and that control units deliver information into the system for further action by the process (perhaps through actuators). Causality determines the direction associated with the information flow.

The basic structures of information flow are a cascade and a loop. So far we have only seen some simply interconnected systems, either a single cascade or a single loop, and occasionally a cascade containing a loop like in the two degrees of freedom controller structure. Also it has been always easy to determine the causality between subsystems, and hence the information flow was always easily recognized.

However dealing with a natural system of some complexity, or even an engineered system, it is not always easy to tell how the information flows. Nevertheless, causality can always be inferred from an understanding of the dynamics. Moreover, when the topology of interconnections becomes cluttered, with many interacting loops and cascades mixed, even simply keeping track of what controls what becomes a major undertaking. The rule here is really divide and conquer. First identify the information flow for each interconnection: what is the cause, what is the effect (even this may be hard if the subsystems at the interconnection cannot be isolated from the rest of the system). Then using the resulting block diagram, subsystems in cascade can be studied separately, or in a single block. Any loops will need to be considered as a single unit. This way we can build a hierarchy of levels of systems, until we arrive at the external signals, the free inputs, and the measured outputs. At the end of this process, we can zoom in and out analyzing the system at any one particular level of its subsystems, and thus building our understanding of the overall behavior.

In a process environment, the Fig. 8.3 schema is quite typical, at least at the highest level. There is the main stream of information from the manipulated variables to the measurements gathered by the Data Acquisition System. The process generates signals

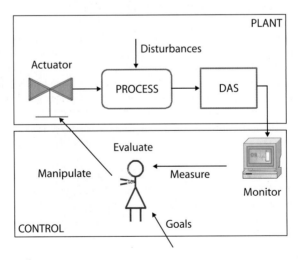

Fig. 8.3. Information flow

which provide the information about its behavior, as influenced by external disturbances. This information is then digested, and presented to the ultimate controller, the human operator who decides where to go next with the system.

This picture at this level of generality even applies to how a country's economy is run. An important collection of operators sets the price of money (the interest rate), which in turn influences the economy, the vital data of which, like trade balances and debt levels and unemployment conditions allow the operators to make further decisions.

In case of a distributed control system, the flow of information is much richer. There are many subsystems intertwined, all with different dynamics, acting over different time scales. Understanding the information flows is essential to come to terms with how control acts in a system.

8.3 Control Goals

No matter at which level we look into a system, the main purpose of a control subsystem, the controller, is to generate signals that eventually (perhaps through human intervention) drive the process it controls, such that the latter behaves in an expected manner. There is always at least one physical interface between the control subsystem and the process, and in a feedback loop there are two, one for sensing and one for actuation.

In some cases the interface is just a direct physical connection. For instance in the Watt's governor (see Fig. 8.1) the displacement of the flyball(s) is mechanically linked to the displacement of the valve position that throttles the steam flow.

In many other instances, the output of the controller exists in a totally different physical domain from the domain in which the input signal to the plant lives, or the sensed variable lives. In such instances a transducer translating signals between physical domains is essential to realize the interconnection. For instance, the controller computes a digital representation of the next input, which must be converted into the appropriate process variable, like the position of a gate, a voltage across a motor winding, or a force or the speed of a motor or the intensity of a light beam and so on. In engineered systems the transduction typically involves the use of electronics, and electrical actuation through motors that then link into the domain of interest. Similarly in the human body, the electrochemical representation of motion in the neural system must be transformed into a muscle force before motion actually occurs. This happens through a chemical process at the neuromuscular junction: a motor neuron's stimulated synapse releases neurotransmitters that bind to a receptor at the muscle fiber which then reacts by contraction.

Based on the objectives and the knowledge about the process (its model, and other relevant information like what are the typical disturbances), as well as the direct measurements, the controller computes the subsequent appropriate actions for the system to react to.

Some common control objectives are

- *Regulation or disturbance rejection* as illustrated by the governor in the steam engine. The goal is to keep the shaft speed at a preset level in spite of disturbances such as a change in the available steam pressure due to changing conditions in the burner. In the irrigation system (see Sect. 3.3) the water level in the supply channel had to be regulated, so that farmers have the right amount of potential energy available to

supply water across their fields. The disturbances are due to the variable water demand on the channel.
- *Tracking*, to follow a reference, like what we saw in the radio telescope example (See Sect. 3.4). The goal is to track a star whose position relative to the observatory is changing over time. External disturbances, like the wind load, must be rejected. Tracking often comes with disturbance rejection requirements as well.
- *Sequencing predetermined procedures* as is required in the start-up or shut-down phase of a process. This happens typically before a process reaches its steady-state, where regulation or tracking becomes the main aim. For instance, before the speed regulator is able to control the speed, the steam must reach a sufficient temperature and pressure. Starting up a steam engine from a cold condition, requires a sequence of events: fill the boiler, check the water level, start the burner, open the air valve, open the fuel valve, initiate ignition, check combustion, reach operational conditions. A similar procedure is to be followed to shut the machine down. Such sequencing, and emergency procedures exist in most processes. In the irrigation system, channels have to be filled at start up. Rain events lead to an emergency shut down. The radio telescope must be stowed under heavy wind conditions to avoid damage. In other applications, like in batch reactors or the simple washing machine or the car wash, sequencing is actually the main control task (see applications described in Chap. 3).
- *Adaptation* such as maintaining overall system behavior despite significant changes, it typically requires large scale adjustment of the control system. Since the dynamics under control change, it is appropriate to also adjust the control subsystem, to ensure that its suggested actions remain appropriate. In the antennae system (Sect. 3.4) the mechanical resonant frequency changes with the pointing angle of the antennae. This affects the allowable bandwidth of the control input, and the controller is adaptively tuned to cater to this.
- *Optimization*, rather than regulating a variable to a specified level, sometimes there is a need to optimize a variable: like in attaining maximal efficiency, or highest power output. This can be seen as regulating the derivative of the variable to zero. Specific control algorithms for such tasks have been developed, like extremum seeking methods. These are closely related to adaptive control and/or learning control methods.
- *Fault detection and process reconfiguration*, as from the monitored process behavior it is feasible to identify alarm conditions, and take action through the control system. In this manner unsafe operating conditions can be avoided either automatically as may be demanded in an emergency, or in advisory capacity to suggest possible actions to an operator. Sometimes, an alarm may require the need to reconfigure how control acts, which is particularly the case when actuators fail. In combination with physical redundancy of actuators, a control system can reconfigure how it implements future control based on the fault detection and/or alarm condition.
- *Supervision* is typical in situations where there are many levels of control, as in the exploitation of a large utility network. The network changes as operating conditions vary, or the network structure itself changes (switches), or components fail. An evaluation of the new condition will typically lead to an assessment of which control objectives remain valid or not, which resources are available for control, and based on this assessment the lower level control subsystems are redirected on how to act. At this level large scale simulation and scenario assessment take center stage in control design.

- *Coordination* is a task normally executed at the highest level of control being typical when there is a clear hierarchy of control subsystems. Coordination ensures that the various subsystems are properly working together. Local control subsystems are directed and provided with references or set-points. When dealing with large scale complex processes, each subprocess will have its own goals, that are aligned with the overall goal. The coordination ensures that the local goals are attained in such a way that the overall process objectives are met. It also assists in start-up and shut-down as well as emergency procedures.
- *Learning* can be achieved from signals gathered during the operation of the controlled system. New information about its behavior can be discovered, and learning or adaptive control techniques can capture this information for later use, like to optimize the controlled response. Learning in itself, in particular when used in feedback leads to very complex system behavior, which is still not well understood, as even in the simplest examples of such control systems chaotic behavior cannot be excluded.

All these different objectives result in very distinct control approaches and techniques: from logic and event-based, to discrete-time controllers, to sophisticated intelligent control systems where ideas from artificial intelligence are key. All these control systems emulate to some extent control as we experience it from our own behavior, just as Norbert Wiener argues in Wiener (1961).

These different flavors of control come with their own analysis and design tools. In modern applications multiple tools are used cooperatively to arrive at an acceptable design solution. Thus far the theory and practice of control engineering has not evolved to the point where there is a standard software "*Control Engineer version 4.13b*" that will guide someone to an appropriate control solution from the problem's inception.

Example

Let us re-consider the steam generator (Fig. 7.8) with the human in-the-loop, as in Fig. 8.4.

The main goal is to produce steam to meet the demand, of course all the while keeping the plant within its proper operational envelope. That means, an appropriate water level, appropriate fuel burning conditions, safe pressure levels, and so on. One aspect of the control unit is to ensure that the water level and water temperature are properly *regulated*.

The operator monitors the overall working of the plant, via a display panel, as in Fig. 8.4. The operator can track the evolution of the system and tune some of its settings. He or she can program the plant to produce a different steam product (changing the pressure, temperature, flow rate) depending on the requirements, either scheduled or as to meet demand. The operator can introduce a *sequential procedure*, to start the boiler up or shut it down, or go into a stand-by mode.

In a more complex operation, the control unit could *supervise* many of the operating conditions and decide automatically about changes in the plant or in the control. For example, based on economic pricing of the fuel, the burners could switch between gas or oil; or the control objective could be switched from regulating the boiler in stand-by to tracking under variable steam load conditions, even *adapting* the subsystem controller that takes care of the water level control, as the dynamics change due to the increased

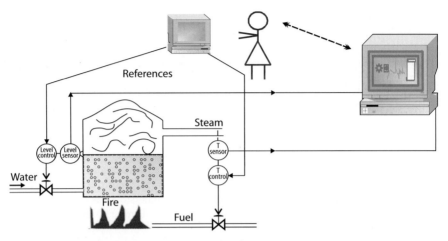

Fig. 8.4. Boiler water level and temperature control with an operator

steam flow. Controllers, subservient to the main controller, can aim at maximal efficiency in the burner, within the limits imposed by environmentally acceptable flue gases.

The control system must be able to *react to faults*. For instance, an emergency shutdown of the burners when the water level is critically low, so as not to physically damage the boiler plant.

Similar situations can be foreseen for all systems described in Chap. 3.

8.4 Open-Loop

As already mentioned, open-loop control refers to control structures where the information flows in a unique sense, without closing a loop. The user determines the parameters of the controller to be applied, the controller generates and applies the control signals to the process, the process dynamically evolves, subject to possible external disturbances and, finally, the user evaluates the results.

In this case, we still have a closed loop, in that the operator closes the loop, but there is no machine-only information loop.

Several open-loop situations can be distinguished.

Sequencing

Consider a washing machine, or a car wash system, or a simple CD player:

- The equipment is well understood. The components, their operation and all the options are clear.
- The process is well-known in advance. Everything has been done before, all events, normal and abnormal are catalogued.
- Disturbances are not really expected, but the process can be engineered to be foolproof (engineers do try).
- The performance requirements are not very strict.

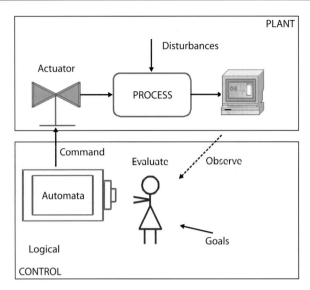

Fig. 8.5. Sequential control

There are many other applications of the same ilk, like starting a car engine, air conditioning, heating a home, operating a video camera. In all these cases, automata provide the appropriate language to deal with the analysis and design of such systems, as shown in Fig. 8.5. Automata are programmed, and then execute a sequence of tasks as events unfold. They may react to specific events (in which case they do use some feedback, and are only open loop in between event times), or simply be time-based in which case they are truly open-loop and entirely oblivious to the unfolding process.

Open-Loop Control

Open-loop means that the manipulated variables are controlled without taking into account the actual evolution of the plant, so there is no information loop.

This is what happens in most master/slave systems. The master generates the control actions (probably in a sophisticated way) and the slave is just applying these signals to the slave system. A key duplicator is a mechanical master-slave system: the "sensor" sweeps over the original key profile and the cutter is tracking this signal on the raw key. Probably we all had the experience that not all the duplicated keys work well at first! If the error is because the cut is too superficial, a second pass can polish the result, but if the error is due to an excessive cut, the created item is useless.

More critical are the master/slave applications familiar from tele-operation. If the application is a surgery, the patient is at risk. If the application is for sampling rocks on Mars a great deal of investment, effort and time is at stake.

The general structure of a computer-based open-loop control system is similar to that of the sequencer. The actions to be applied are often computed in advance, stored in the controller and applied at a later time. The main drawback is indeed the lack of (immediate) feedback. If the actual system's behavior is not as expected, it deviates from the model used to compute the control actions, or if there are some disturbances not taken into account in this computation, the actual system's response will not match expectations.

On the other hand, if the operational conditions are precisely those used to generate the control signals, the system will perform without any error. The controller does not need the existence of an error to generate the control, as it is the case in feedback control. The overall system is easier to implement. Most robots used in manufacturing operate in open loop.

Feedforward

The control signals can be generated on-line in real-time based on information gathered from sources other than the system itself. *Feedforward* control applies to an open-loop control structure where the control action is computed based on measurements of disturbances or references, as shown in Fig. 8.6.

Let us consider again the boiler system. In this case, we focus our attention to the level regulation in the water tank. As soon as a user opens a valve to get more steam, we do not need to wait until the water level decreases. We know that steam consumption requires more water and hence in response to increased steam demand the controller opens the water inlet valve (and may also increase the burner's heat output).

Sometimes like in multivariable systems, the disturbances on one controlled system are actually created by the actions from another control subsystem that controls another variable that interacts with the first control subsystem. Again in the boiler example, if we try to control simultaneously the water level and the temperature we know that a desired increment in the temperature will need an increment in the heat output of the burner. This modifies the equilibrium water/steam, the pressure and also the

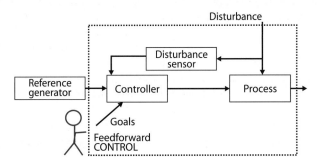

Fig. 8.6. Feedforward control

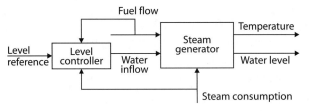

Fig. 8.7. Feedforward control of water level

water level. This interaction must be taken into account to reduce undesirable actions from the water inlet valve. These ideas are illustrated in Fig. 8.7.

Similarly, when tracking a reference signal we are able to know this reference signal in advance so we may exploit this preview knowledge to ensure that the system response is well-aligned with the reference at all times. We use this strategy when driving a car, we look ahead not in the rear view mirror to steer the car. Driving a car by feedback only would equate to driving using information only from the rear view mirror and other instruments in the car.

Even so, feedforward is rarely implemented without a feedback loop of some sort, alarms must be monitored, and goal attainment is to be verified.

Quality Control

A special case of an open-loop control system goes by the name of *quality control*. In this methodology, quite common in manufacturing industries, measurements, evaluation and corrective actions are taken after the completion of the entire process.

For instance, consider the quality control of the ceramic tiles in Chap. 3. The procedure involves

- measuring the tile characteristics (size, defaults, appearance, ...);
- classifying the tiles against predefined standards;
- statistical analysis of the data, leading to a proposal of future actions, perhaps changing process variables or using new set points in the control subsystem;
- reporting the findings to management.

In general, these activities are performed off-line, and often the reaction time is actually longer than the production time of a single batch of product.

In order to gain more from the quality control observations, some of the activities could be fast tracked, to approach real-time feedback. Often this uses heuristics and/or rule-based actions on partial measurement results. In this way corrective actions can be applied in a shorter time frame. The investment is worth it if there is a sufficient economic return because of improved quality or throughput in the production line.

> **Open-Loop Control**
>
> In open-loop control information flows in one direction, without feedback. It works well if:
>
> - The process is well-characterized.
> - Disturbances are not important or they can be adequately measured and counteracted.
> - Performance requirements are not very stringent.
>
> It is certainly not appropriate if:
>
> - There is uncertainty about the process behavior.
> - There are unknown and/or unmeasurable disturbances.
>
> The dynamic behavior of the whole system is readily deduced from the dynamics of the subsystems.

8.5 Closed-Loop

Closed-loop control uses response information to determine the input. Feedback is advised when some or all of the following conditions apply:

- Disturbances are unavoidable.
- System response information is available. This is the *sine qua non* for feedback.
- Performance requirements are demanding.
- System dynamics are poorly known, or subject to large uncertainty, or variable over time.

In feedback, Fig. 8.8, control signals are based on all available information, process knowledge, which includes a model for the uncertainty about the process knowledge, disturbances information (measurements, and/or models), objectives to be satisfied (reference tracking, regulation and so on) and last but not least the actual system response measurements.

What infrastructure is required for feedback? Obviously, there is always a cost associated with feedback. Instrumentation is essential, both for measurement and actuation. The answer to the above question really depends on the overall requirements. The main issue is simply this, the feedback must be able to decide if the requirements are met or not, and have enough action capacity to enforce the requirements.

Sometimes simply the consideration of the objective is all that is required for feedback. For instance, the room temperature regulation by a thermostat is simple: heating on when too cold, heating off when too hot. The thermostat receives a temperature reading that can be rather crude: temperature below or above or inside the reference band is all that is required.

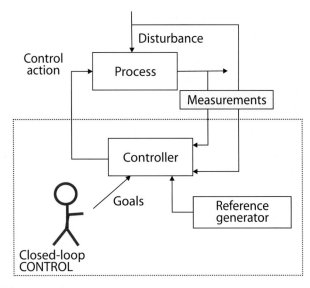

Fig. 8.8. Closed-loop control

In general though, more and more precise information leads to better feedback. There is of course a rule of diminishing returns. Importantly, all measurement information that feedback could possibly need is captured in the notion of a state for the system. A system state is defined as sufficient information at a given time to be able to predict the system evolution given the present and the future inputs. If we can find a *minimal* (smallest dimension) state, that is enough for feedback. There is simply no more information that can be gathered about the system. Hence *state information* and *state feedback* is the most we ever can do.

It sounds great, but state feedback requires in general a lot of instrumentation and a lot of communication capacity to transmit the data. More often than not, state feedback based on full state measurements is prohibitively expensive or cannot be implemented because the state cannot be measured (e.g. try to measure the temperature *inside* a bauxite smelter).

In such situations, and knowing how powerful the state information actually is, the closed-loop control may consist of two separate systems: first a system that takes all available information (measurement, model etc…) and produces from this a (best) guess or estimate for the state (or any desirable internal variable) and next a controller subsystem that uses the estimate of the state to produce the next input value. The first subsystem, from measurement and model to state, is also called a *virtual sensor*, nowadays incorporating many features related to the quality of the measured signal itself. The celebrated *Kalman filter* is such a virtual sensor; this is discussed in Sect. 9.2.

An alternative, using partial state information, is discussed in the next section.

> **Closed-Loop Control**
>
> The basic control loop requires a sensor, an actuator and a controller. Their interaction with the plant to be controlled defines the global behavior of the controlled system.
>   Closed loop control is essential when:
>
> - the process has to operate in an unstable regime;
> - there is large uncertainty about the process behavior;
> - there are unknown disturbances.
>
> Feedback must be tolerant of some level of errors, otherwise it can never act. Feedback must be provided with measurements that either directly indicate, or allow the inference of how well the control objective is met.
>   Feedback may be inappropriate when:
>
> - the required instrumentation is too expensive or does not exist;
> - there is no plant uncertainty, nor disturbance;
> - feedback design cannot be verified.

8.6 Other Control Structures

Besides feedforward and feedback control many other control structures are used in practice, but they can be seen as a combination, or perhaps a repeated combination of these two basic concepts.

Using a few simple examples, we review some of these other control structures.

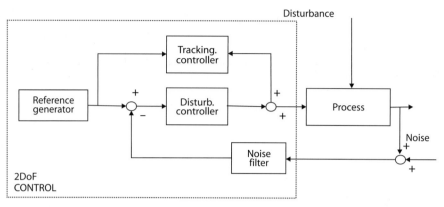

Fig. 8.9. 2DoF (two degrees of freedom) control

8.6.1 2DoF Control

The stability properties of a controlled system are determined by the control loop(s). When stability is of concern, and there are also disturbances acting on the system, and/or there are tracking requirements, a single loop control strategy does not provide enough design freedom to achieve all objectives. Because signals can be shaped by systems, the tracking performance clearly depends not only on the loop but also on any system in cascade. This reasoning leads to the very common control structure with two degrees of freedom, as represented in Fig. 8.9.

The antenna system described in Chap. 3 is a good example of such a controlled system. The antenna should reject the effect of external forces, such as the wind. Also, the antenna must point to an object in the sky that moves (star, satellite). The servosystem is a two-degree-of-freedom controller that ensures sufficient damping of wind effects and ensures at the same time accurate tracking.

8.6.2 Cascade Control

Full state information may be unavailable, or difficult to use at once. In cascade control, successive control loops are used, each using a single measurement and a single actuated variable. The output of the primary controller is an input to the secondary and so on. A judicious choice of how to pair variables and the ordering of the multiple loops can lead to a very efficient control implementation without the need for a full state feedback control.

Again the servosystem for the antennae is a good example. In the end the antenna needs to point to the right position. The position error can be used to set the reference for the motor, but the motor speed and its torque are also important variables (that describe in part how the system works, and they are part of the state of the system). The motor speed can be measured, and so can the current (which is related to the torque) and these can then be used with advantage in controlling the overall behavior. See Fig. 7.3. In this way a multiloop or cascade control system, as depicted in Fig. 8.10, can be implemented.

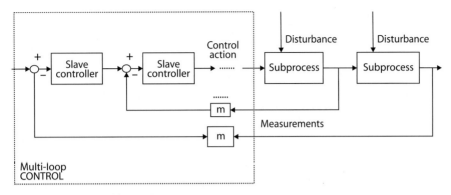

Fig. 8.10. Cascade control

The inner loop will regulate the motor current, effectively ensuring that the motor behaves as a torque source. It eliminates load disturbances. Then, the speed feedback will provide damping and stability. In the outer loop, the position controller (the master controller) will take care of the final position of the shaft. A nice aspect of why this works so well, comes from the fact that the loops have different reaction times. The inner loop is fast, about ten times as fast as the damping loop, which is itself a factor of ten faster than the outer position loop. In bandwidth terms, the outer loop has the smallest bandwidth, the damping loop's bandwidth is roughly ten times larger, and the inner loop another factor of ten larger again.

Thus, the main concept behind the cascade control is: *feedback any information in the plant as soon as it is available.*

8.6.3 Selective Control

Let us consider a waste water treatment plant. All pollution must be removed, and the cleaned effluent should meet all required safety and environmental regulations: its temperature, pH, biological contamination and turbidity must be regulated.

Under normal operational conditions, the pollution controllers determine the maximal effluent flow rate, so that the effluent meets expectations. However, in an emergency, like flood conditions, the controller will temporarily "forget" about pollution control and will determine the outflow to avoid a massive plant failure that may be caused by overflowing storage tanks.

In combustion control dealing with both fuel and air flow, safety will always override normal combustion control to generate heat for the boiler.

Also, to avoid explosive mixtures in a burner, typically air should be in excess. Therefore if there is an increment in power demand, the air flow is increased in advance of the fuel flow. In the opposite direction, if the purpose is to reduce power, the fuel flow is reduced first, followed by the air flow. Thus depending on the required plant evolution either air or fuel tracks the other one.

An example of selective control is also illustrated in Chap. 3, talking about the purpose of exercise (see Fig. 3.26) in deciding the insulin regime for a diabetic.

8.6.4 Inverse Response Systems

As already mentioned, one of the basic features of feedback control is reaction: first an error is detected, then follows the (re)action.

In many systems the error response at the onset is in the same direction it will settle into the future, as long as there are no corrective actions applied. Unfortunately, there are systems where this is not the case. The initial response goes in the opposite direction of where the final response will evolve towards.

Examples are systems with a pure time delay. Actually these do not have any response at all in the initial stage.

There are also the so-called non-minimum phase processes[2]. In these circumstances, the initial response will mislead the feedback controller about what to do. In fact, these systems are much more difficult to control and there are inherent limitations in what control can actually do, no matter how hard we try.

A typical example is the water level control in a boiler. Clearly to keep the water level at a given reference point, the inlet water flow is to be increased if this level drops below the reference. Under normal operation, when the outlet steam flow is increased, the boiler pressure reduces. As a consequence, more water goes to boil, creating more air bubbles inside the water mass, and this has the effect of expanding the apparent water volume in the boiler. The water level sensor observes an increase in the water level. This would require a reduction in the flow of the inlet water, and yet the opposite is required. A typical response is shown in Fig. 8.11. An increment on the steam outlet flow at $t = 6$ s leads eventually to a final reduction of the water level of around 4 cm, but in the first two seconds, the water level actually increased by more than 1 cm. A similar problem occurs when increasing the inlet flow, as this relatively cold water enters the boiler this cools the water mass in the boiler, bubbles disappear and the water level goes down.

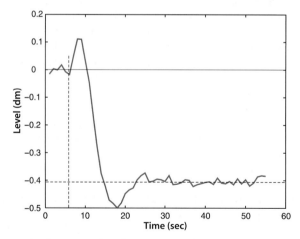

Fig. 8.11. Boiler water level response when the steam outlet flow increases

---

[2] In the context of linear systems, a non-minimum phase system has a transfer function that has zeros with a positive real part i.e. there are unbounded inputs for which the system has no response at all.

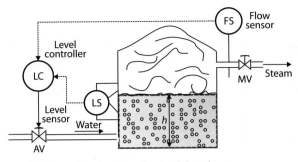

Fig. 8.12. Boiler water level control under steam flow load disturbances

For these kinds of systems feedforward control may help. For instance, in the previous example, if a steam flowmeter is installed, the water level controller can know in advance the forecasted effect of steam flow variations (due to changes in the manual valve MV) and modify the water inflow accordingly (acting on the automatic valve AV) preempting a level error reaction. The measured changes in the steam flow will be introduced as changes in the water flow reference. Nevertheless, as there will be always some difference between the steam and water flows, a final more slowly acting feedback compensation will be required. A typical control lay-out is depicted in Fig. 8.12.

8.7 Distributed and Hierarchical Control

In nature and in engineered systems, many system properties are actually distributed over space, and not characterized by a single or a few measurements. Consider, for instance, the temperature control of the human body, as depicted in Fig. 8.13. First the required body temperature is not the same for all parts of the body. Our heart and brain require a different temperature from our feet and fingers. Moreover many different controlling mechanisms are in operation. In the end though, all these temperature control subsystems are interconnected and must work together to achieve a common purpose. Also there are many different temperatures: skin, muscles, viscera, the brain. There are temperature sensors everywhere as well as end effectors, to react. Some actions are elaborated locally (spinal cord) and some others go through the brain for decision making.

In human made control systems the same rule applies. In a complex system, although the global goal could be the same (optimize the consumption, maximize the benefits) each part of the system has its own control options and local goals. Nevertheless, like in the human body, the communication channels will allow a coordination of controllers to achieve the global goal.

Thus, in a distributed control system we may find simple on/off controllers, automata, simple or sophisticated local controllers and, a lot of information exchange between subsystems to coordinate all the local activities.

Remember the manufacturing of tiles as introduced in Chap. 3. The real objective is to produce as many quality tiles as possible at a low cost with minimal pollution. The system is totally distributed with a number of local controllers for different subprocesses. The command signals for all the subprocesses must be coordinated to achieve the overall objective.

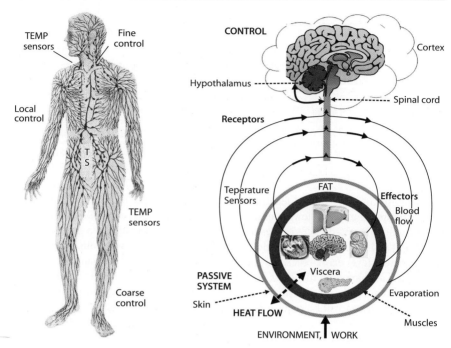

Fig. 8.13. Body temperature control system

Nowadays many distributed control solutions are available. In fact this aspect of control is spurred on by the emerging technology, of sensor/actuator networks. The designer can select components, a communication network technology, appropriate computer control hardware, and appropriate software, and design the system to act in unison. The distributed control task can be conceived as a decentralized system (loosely cooperating local controllers) or as a hierarchy of different levels of control or a mixture. To know what is the actual information topology as well as what information is available for which actions, is in fact an untractable problem in its full generality. Heuristics, experience, and expert design are called upon to come to a robust well-performing and coordinated system.

> **CCC: Communication, Computation and Control**
>
> The new information era is grounded on three pillars:
>
> - Computers to deal with information (hardware and software)
> - Communication to make information accessible (channels, emitters and receivers)
> - Control to design what to do with information (algorithms)
>
> In this way, networked control, as applied to distributed control problems, is based on the technological advances in communication, computation and control. Sensors and actuators equipped with a (wireless) networking enable feedback. This feedback must be designed to utilize the (always) limited resources of communication and computation to achieve the maximum benefit in system behavior. Without control design, communication and computation leads to a glut of data without information content.

8.8 Integrated Process and Control Design

In the past, it was normally the case that a control subsystem was appended to an existing process to improve its dynamic behavior. As a result, the only option for the controller is to "select" the best process outputs, and inputs from whatever is available. If the process is not designed to accept a particular well-suited input or the variable of real interest is simply not accessible these opportunities are lost to the controller. Many of these constraints are avoidable when the process and controller are conceived together.

Recently a new paradigm has emerged: integrated design of process and control. There is an important interplay between process and control design. Small changes in the conception of the plant can make its control much easier and hence stronger performance may be achieved. In fact, any change in the process design influences the process dynamics and hence how its control is approached.

Conservative process designs are often an obstacle to achieve better results in controlled operation. For instance, an aircraft designed for ease of manoeuvring at high speed may well be unstable at low speed, and hence cannot be piloted. In the aircraft design phase this is then rejected as an unsuitable design. But if the control is designed at the same time, it could be feasible to design a feedback controller that allows the pilot to fly the plane both at high and low speed, because the feedback controller can deal with the instability at low speed. The result is an improved overall performance. A similar example of such integrated design was briefly touched upon in Sect. 7.7. Intuitively manoeuvrability and stability are contradictory concepts in mechanical systems, requiring strong performance on both accounts really demands an integrated control/process design approach.

In exploiting the use of an exothermic chemical reactor, it is often the case that the best outcome is obtained if the reactor is operating around an open-loop unstable equilibrium. An appropriate control system (with a careful safety net, ensuring that there is a way of coping with a failure in the control subsystem) will allow one to exploit this possibility.

Let us conclude our discussion of controller-process design with two more illustrative and classic examples from process control.

8.8.1 Scaling the Process and Its Control

One of the most challenging problems in process control has been the control of the pH of a large volume/flow of say water. This is mainly due to the extreme range of hydrogen concentrations (the scale extends over 14 orders of magnitude) and the extreme sensitivity with which we can keep track of hydrogen concentrations in water. To make matters worse, typically the point for pH regulation corresponds to maximal sensitivity. Extreme sensitivity of the response to the input at the point of regulation is not a good situation to begin control with.

As a result, it is almost impossible to get the right pH in a large tank, it would take forever to control. A practical solution is to scale the process and the control, as shown in Fig. 8.14

In option **a**, the effect of the disturbance (acid flow into the tank) is observed by the pH sensor (probably with some inherent delay) and the pH controller commands changes in a base flow to neutralize the acid. In option **b**, the same is done in two stages.

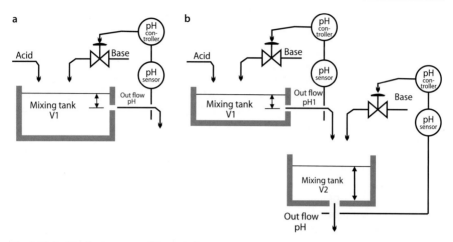

Fig. 8.14. Scaling the process and its control

Dealing with a strong acid (pH < 3) and a strong base (pH > 11), a large control tank requires the patience of Job to settle, as well as an impossibly large actuator with an impossible fine resolution. Not feasible.

Using option **b**, using smaller vessels, in the first tank (a faster and) coarse regulation is achieved using a coarse grained large actuator. In the second tank, the final goal is reached using a smaller actuator with a finer resolution.

Clearly, the process needs to be designed with control in mind. It is not possible to achieve good regulation with a single tank process.

8.8.2 Process Redesign

Similar to the example of the manoeuvrability of the aircraft, we illustrate the design of a typical distillation process unit with two subsystems. The distillation process is split into two parts. If they are conceived independently, the heating/cooling systems are designed and controlled separately, as shown in the upper part of Fig. 8.15a. On the other hand, the system's efficiency is higher and the control easier if both systems are designed jointly and the control of the cooling subsystems is integrated, as shown in part b of the same figure.

> **Intertwining Process and Control**
>
> When designing a new system, it pays to consider the co-design of control and process. The benefits are large when:
>
> - All performance expectations are expressed in advance;
> - Control and process are subsystems that are co-designed to meet the performance expectations;
> - Feedback is explicitly catered to;
> - Maximal design freedom in both control and process subsystems is used to, explore performance limits.

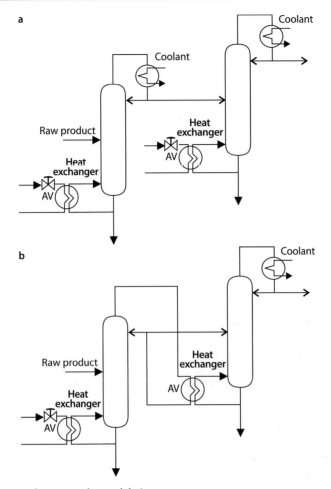

Fig. 8.15. Integrated process and control design

8.9 Comments and Further Reading

In this chapter the basic concepts related to control have been summarized. Feedback plays an important role, but clearly it is only part of the story.

It is perhaps timely to remind ourselves that in feedback we are in the first instance designing with models, not the real thing. Therefore it is important to realize that models are in some sense convenient lies, and we cannot get carried away with the performance of our models and simulations. As always, if it looks too good to be true it probably is.

There are a large number of texts dealing with various design and optimization problems in control. Books dealing in-depth with the process-control interaction typically are about transducers. These tend to be specialized for particular domains, electrical or sound or piezo-electric transducers for example, or deal with a particular application domain. Indeed domain knowledge is extremely important in this context.

Kawaguchi and Ueyama (1989) deals with control in the steel industry. A substantial piece of work just about how the body regulates temperature is Hoydas and Ring (1982). Control and a good understanding of the process must go hand-in-hand to achieve the best possible outcomes, e.g. McMillan and Cameron (2005) deals exclusively with pH control. This is also observed by considering the many different commercial control service providers, they tend to cluster their services in a particular area or domain of expertise. The same can be said for how control design is often taught in disparate manners in mechanical, electrical, chemical or biomedical engineering curricula. Perhaps it should not be so.

A substantial text dealing with control system design in general is Goodwin et al. (2001). In the process control industry Shinskey (1996) is very popular. A book advocating the integrated design approach is Erickson (1999). Multi-loop design is treated in Albertos and Sala (2004) and Skogestad and Postlethwaite (1996). The list is truly immense and the interested reader will easily find additional pointers.

Chapter 9 — Control Subsystem Components

9.1 Introduction and Motivation
9.2 Sensors and Data Acquisition Systems
9.3 The Controller
9.4 Computer-Based Controllers
9.5 Actuators
9.6 Comments and Further Reading

Chapter 9

Control Subsystem Components

> *The whole is more than the sum of its parts.*
> Aristotle
> *The whole is less than the sum of its parts.*
> J. W. Gibbs, Jr.
> **?**

9.1 Introduction and Motivation

In a control context, we have to agree with both Aristotle[1] and Gibbs[2]. Aristotle is right, as there is indeed little point in building a control system unless we gain something from combining the parts. After all the raison d'être of control is to enhance the behavior of the system. Though Gibbs is equally right in that by the very act of interconnecting subsystems we actually have imposed constraints, hence lost some freedom and made the whole simpler to understand. In this way, they are both right.

The illustration on the previous page captures the main components of a control subsystem by the operation of decanting water into a glass. The whole process is coordinated. The two hands are aptly guided by the brain, using feedback signals from the eyes and the muscles.

The basic control loop is composed of: the process to be controlled (the pouring of water from the pitcher into the glass), the actuators (the muscles acting on arms, body and eye position), the sensor system providing information about the process variables (the eyes provide visual information about water flow, the pitcher and glass positions, but also the body and legs' positions to perform the activity, force information comes from the muscles), and the controller (the brain generates the signals to command the actuators, but also local controllers instinctively react if, for instance, some water drops on the left hand).

In human-made control systems, all these components appear as well, typically supported by different technologies. In a control loop (Fig. 9.1), so far we have always distinguished two subsystems, the process and the control. The operator interacts with the control loop, typically by acting on the controller subsystem directly, (see e.g. Fig. 9.2b). More realistically a control loop looks more like Fig. 9.2a. where the separation between control and process subsystem(s) is less clear.

Presently most controllers are digitally implemented using microprocessors or general computer units being composed of (see Fig. 9.3)

- *Sensors or Data Acquisition Systems* (DAS). They measure the variables of interest, and often present a digital output by incorporating an *Analog to Digital Converter*

---

[1] Aristotle, 384–322 BC, Greek philosopher, whose work has greatly influenced scientific thinking through the ages.
[2] Josiah Willard Gibbs Jr., 1839–1903, brilliant thinker, who earned the first Ph.D. in engineering in the USA in 1863 for his now breakthrough work on the foundations of thermodynamics.

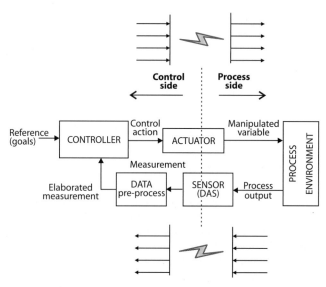

Fig. 9.1. Components in a basic control loop: process and control sides

(ADC). Their input (and their internal working) is determined by the process variables to be measured. The sensor, in particular when its own dynamics are important, is often considered as part of the system to be controlled.
- *Data Pre-processors or Data Signal Processors* (DSP). These digital computing units perform scaling, filtering smoothing or any other preprocessing function solely based on the digital output from the DAS. This function could equally be performed by the computer implementing the control algorithm. In other instances the DSP is integrated with the DAS.
- *The controller algorithm* The "brain" of the controller subsystem. It orchestrates a series of activities, not only the control computation, but also alarm treatment, fault detection, supervision and coordination (through communication with the external world) as required. These various tasks are discussed in more detail in the next chapter.
- *Actuators* The actuators are the brawn of the control subsystem. Its task is to deliver the input to the process. Its front-end, or input side, may contain a Digital to Analog Converter (DAC), otherwise, a special transducer has to be added as an interface between the control output and the actuator input. Often actuators are considered as part of the process to be controlled, in particular because actuators have their own dynamics and more importantly their own limitations.
- *Communications network*. The sensors, actuators and controller are not necessarily in the same location. Typically the data from the sensor(s) are put onto a communication network. The controller(s) and actuator(s) also have access to the same network, and all of them are uniquely identified on the network. Think of it in this way: they all have a mobile phone, and are able to call, or better, to message each other individually or in groups. All data exchange happens via the messages transmitted on the network.

9.1 · Introduction and Motivation

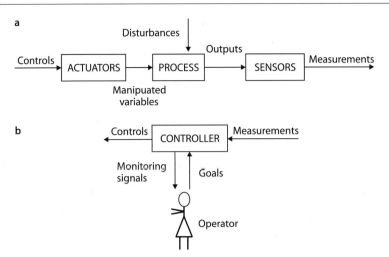

Fig. 9.2. Control cascade: **a** process side; **b** control side

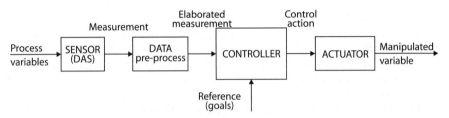

Fig. 9.3. Components of the control side

In broad terms, the process side deals with analog, continuous time signals whereas the control side deals with discrete time, digitally quantized signals.

The communication channel allows for a spatially distributed system implementation. The control computer does not have to be located with the process; whereas actuators and sensors are always located with the process. Also, from this point of view, it makes a lot of sense to consider the sensors and the actuators as a part of the extended process.

A control system implemented across a communication network has a hybrid time and hybrid signal character. This raises some interesting questions.

1. As the process and the control must communicate, an analog to digital and digital to analog interface is always required. Dynamic range (total range of values to be represented) as well as the accuracy (the number of levels to be distinguished) are critical here, and form part of the design. Both range and accuracy will limit the performance of the controlled system and both place demands on the communication system in terms of number of messages, or size of messages required to communicate.
2. As digital systems work using quantized signals and the process is using analog signals, a hybrid analysis (hybrid as in analog and discrete together) is required. How well do the quantized signals actually represent the analog signal?

3. As the final goal has to do with the behavior of the process, a discrete time analysis of the controlled system is not enough because the intersampling behavior (what happens in between the times the controller intervenes with the plant) also counts. Over the time interval that the controller uses to compute the next control signal and until the actuator implements this new signal, the process is in open loop.

Despite the obvious differences between discrete and continuous time, between continuously ranged or quantized valued signals, we will treat them very similarly and as much as possible in a unified way.

The advantage of the digital world is the flexibility to integrate components, and to scale up. Even the communication network appears from a user point of view as part of the digital world, and yet at its core it is very much an analog signal world.

9.2 Sensors and Data Acquisition Systems

A sensor is a device (it could be a biological organ) that detects or measures a signal in a process. The main feature of any sensor is that it must minimally impact the process: it measures and represents the process behavior with high fidelity in real time, without altering the process behavior. (Invariably though any measurement does extract some energy from the process, and hence does influence somewhat the process behavior.)

The detected signal is then converted into a mechanical, electrical or digital signal so that it can serve as an input to the control subsystem (or as a valid message into the communication channel).

This conversion is the role of a transducer. Usually, sensors and transducers are integrated, and often integrated with some digital signal processing capability as well and the whole is denoted as the data acquisition system.

At its core a sensor exploits a repeatable, well-known physical relationship between its analog output (voltage, current, position) and its analog input, the process signal that is measured.

Nowadays in an engineered system, that basically means that the output of the sensor consists of a digital message with preferably three pieces of information: the sensor identity (with its calibration data), the signal value, and the measurement time.

In order to get the time, the sensor of course needs a clock. As time is of the essence in dynamics, this time must be right, and over the entire control and communication network the clocks must be (sufficiently) synchronized (otherwise rather strange things can happen).

The advantage of living in a purely analog world, e.g. remember the steam engine with the governor, is that there is no need for a clock. Time is implicit in the entire behavior. We, the external observer, only impose time to describe the evolution of the system over time.

The sensor identity is particularly important information, because sensor values by themselves are meaningless. Sensors do need calibration to make their output intelligible. Indeed, the relationship between the observed variable and the sensor output is always nonlinear (no sensor has infinite dynamic range and infinite precision, both of which are necessary for linearity). Moreover, typically the relationship between input and output is not static, it is dynamic. Worse still, the relationship varies with operating conditions (temperature) and age. Sensors wear out, develop errors and so on. So in order to make a sensor really work, a lot of additional circuitry and monitoring must take place. The term sensor system is well-deserved.

In order to interpret the measured signal, its engineering units, the scale, the sensitivity, in general the entire sensor configuration must be known. This configuration process may be simple when only a few sensors are involved, but it is cumbersome and requires attention to detail when large numbers of many different types of sensors get connected into a multichannel measurement systems.

Setting aside, the basic activity of measuring a variable, which is very particular for each sensor (such as the thermocouple or a tachometer, a strain gauge, a flow meter or a Ph-meter), the desirable features of a modern sensor or *data acquisition system* are:

- convert the measured variable into the appropriate physical domain,
- *ADC* convert from analog to digital,
- measure the whole range of interest,
- present a linear relationship,
- reduce measurement time, delay and/or dynamic effects,
- deliver the required accuracy,
- avoid errors (drift, bias, noise),
- detect malfunctions and signal fault conditions,
- execute self maintenance,
- alert for the need of recalibration,
- have an appropriate communication interface, that allows a variety of external user(s), and other devices, to interrogate and/or calibrate the sensor.

Calibration and in general the conversion that occurs in the sensors are critically important to interpret their digital output. A markup language is being developed to standardize information that is required about sensors in order to be able to interpret the data meaningfully. SensorML is one such sensor definition standard under development[3].

In a feedback context, sensors play a critical role.

- Any measurement error will mislead the control. Quantization errors, out of range errors and dynamic response of the sensor will limit control performance.
- Sensors accuracy imposes limits on control action accuracy.
- The loop behavior is influenced by the sensor dynamics (delays, additional time constants).

The impact of these issues can be reduced somewhat, but not eliminated, using redundancy. Sensor redundancy can be achieved either by installing additional sensors (hardware), or using soft sensors (software) as part of the control algorithm that combine the system model with the measurements to improve on the latter. In the same vein, feedforward control can side step some of the issues.

---

[3] The Bureau of Meteorology of the USA is developing this standard to coordinate all the sensor information across its domain of interest: weather forecasting. See also *http://www.opengeospatial.org/standards/sensorml*.

Control–Sensors Interplay

Sensor behavior limits control and feedback performance.
Quality feedback demands quality sensors.

Nowadays, plug-and-play sensors and so-called *smart sensors* provide many of the above features and do indeed a lot more than purely measuring. The sensor system is a small computer and communication device, incorporating additional circuitry consisting of amplifiers, data converters, micro processors, firmware, and some form of nonvolatile memory. As they are computer processor-based devices, such sensors can automate the removal of nonlinearities and offset and gain errors from raw sensor readings, thus eliminating the need of central post-processing at the control processor. The calibration data on a smart sensor can also be stored locally so that the module as a whole can be moved and often reused without the need for manual recalibration. Some smart sensors systems come as system-on-a-chip, i.e. they are fully integrated in a single silicon electronic device and can be easily added as embedded hardware in a larger system.

The diagram on Fig. 9.4 shows the sensing device (a thermocouple, a resistor bridge) delivering an analog signal, an ID block, including digital data about the sensing parameters, calibration and sensor placement, an analog-to-digital mixer/converter, and a data preprocessor preparing the measurement to be easily used and handled.

In remote data sensing and networked-based control systems, a wireless-equipped smart sensor can perform local processing of the raw data and then send the processed data through the network using an appropriate communication protocol. In order to facilitate this technology, a number of interface standards are being developed.

9.2.1 Transducers

As indicated there must be an interface between the control subsystem and the process that converts the control output signal to a process input signal. Similarly the plant output measurement has to be transformed into a signal that can be interpreted by the controller.

In some cases the interface is immediate, like for instance in Watt's governor.

More typically, the interface consists of a transducer that transform signals between different domains as required. The output of the controller is for instance an electrical quantity such as a current or a voltage or just a number represented by a computer. This signal needs to be converted into the physical domain of the input to the process.

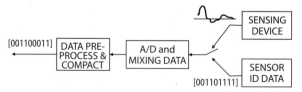

Fig. 9.4. Components of a smart sensor

The typical transduction path from controller to plant is from a (digital) number (inside the computer) to an electrical signal (a voltage or a current), to a mechanical movement (a motor), to the domain of interest. For example, to change the heat output in a gas furnace we may have the following chain of subsystems. A digital number (controller output) is converted into an electrical signal using a digital to analog converter (DAC). This electrical quantity is used to drive an electric motor or actuator. The motor turns a shaft that rotates a valve. The latter modifies the gas flow which in turn adjusts the heat output in the gas burner.

Similarly a transducer is required to link the temperature in the burner to the digital controller. The temperature in the burner may be measured using a thermocouple, a junction of two metals whose temperature creates an electric voltage across the junction. The latter is measured, and translated through an ADC so that the control computer can interpret it, and compare the actual temperature with the desired temperature setting and provide for an appropriate input setting for the gas flow.

A force can be measured using a spring, the elongation of which is proportional to the size of the force (Hooke's law). Equally a strain gauge could be used, on the principle that a force modifies the resistance of a resistor subject to this force. The latter can be observed as a change in voltage.

A pressure variation can be picked up using a electrostatically charged capacitor with a spring loaded capacitor plate. The pressure exerts a force on the capacitor plate, and its deviation leads to a change in the voltage across the capacitor. This principle is at the heart of electrostet microphones. Variations in plate size, spring loading and charge density leads to a variety of pressure to voltage transducers capable of dealing with a great variety of measurement problems. The inertia of the plate and spring indicates that there is a certain dynamic response that is to be minimized in order to get a high fidelity representation of the pressure to be measured.

Biology provides some really intriguing examples of transducers. For example the hair cells in the inner ear, vibrate as a consequence of a sound wave acting on a membrane. This vibration in turn mechanically opens and closes pores in a membrane allowing ions to flow, and create an electrochemical stimulation of the auditory nerve cell. The response of the transducer is in itself regulated through feedback to ensure that only few haircells respond to a particular tone in the sound wave. In this manner every sound is immediately coded into tones, as only those haircells that represent the tones in the sound react. Moreover the intensity of each tone in the sound is coded by the intensity of the response at the haircell.

9.2.2 Soft Sensors

In some instances, the required variable cannot be directly measured through a hardware sensor, but its value may be inferred from knowledge of the dynamics and other signals that can be measured using hardware sensing. This is a so-called *soft sensor* or *virtual sensor*.

Figure 9.5 illustrates the concept: based on information gathered from the process, including manipulated inputs and disturbances, as well as a model of the process, some or all of the internal variables may be estimated.

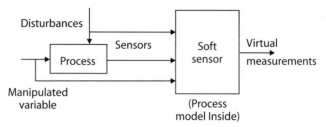

Fig. 9.5. Information processing-based virtual sensor

The idea of soft or virtual sensing is based on the notion of *observability* and uses ideas from filter theory. Observability is the concept that from knowledge of the process model and the mere observation of the external signals of a system, i.e. the inputs and outputs it is possible to reconstruct the state. It is one of the key notions developed in systems theory. Observability is a remarkably generic property, in that almost all (linear and nonlinear) systems possess this property, but it is not universal, and there are degrees of observability: some variables are easier to infer than others.

Kalman Filter

The most successful idea in virtual or soft sensing to date is the so-called Kalman filter. The latter pervades most of the control and signal literature, and most of its implementations, since Kalman[4] wrote his seminal paper in 1960.

A Kalman filter or a variant, that is able to estimate the full state of a model, requires:

- A model for the dynamics of the process, in particular a state description, including how the input acts on the state, and how the state relates to the output. The model does not have to be precise, an estimate for the model error is necessary.
- A model for the measurement system, in particular a model for the statistical properties of the measurement errors.
- A sufficiently rich output signal that can be measured (observability must hold).
- All input and output signals must be measured.
- A reasonable initial guess for the state.

The enormous success of the Kalman filter is due to the fact that the requirements on the modeling side are actually quite minimal. A simple, reasonable model, with an estimate of what the model error may be, is sufficient. Of course the better the model, the more we can expect and demand from the soft sensor.

---

[4] Rudolph Kalman, 1930–, born in Hungary, studied at MIT obtaining his masters, and completed his Ph.D. studies in 1957 at Columbia University. He is famous for the exposition of the filter, now called Kalman filter. The first Kalman filter was implemented by Stanley Schmidt at NASA and used in the Apollo space program the same year Kalman published his seminal paper. Since then, Kalman filters are the systems theoretic concept with the greatest patent impact ever.

9.2.3 Communication and Networking

Most components in a control system are networked, exchanging information through a communication network. There are many ways to implement this network, providing different levels of security, speed, concurrency and robustness. There are wired, wireless and mixed radio with wired control networks. A number of standards exist, and more are being developed to cope with new technologies and requirements. This is the realm of Supervisory Control And Data Acquisition (SCADA) systems.

The trend is towards integrating data, communication and computing networks so that the administrative side of a process is linked into the day-to-day control side of the plant. For example, in a modern ceramic tile factory, the information from the sensors used in controlling the burners in the kiln is also linked into the accounting software for the plant operations. In the irrigation system the radio network used across the system for SCADA can be integrated with a broadband internet access network for the remote communities serviced with irrigation water. (See also Chap. 3.)

9.2.4 Sensor and Actuator Networks

Electronic developments, in particular the capacity to integrate radio on a chip, and build systems-on-a-chip have enabled the development of *sensor dust*, (see Fig. 9.6). These are tiny devices capable of sensing different variables (temperature, proximity, and so on) also able to form an ad-hoc communication network to transmit sensed data as necessary. This development is driving home and building automation, and is identified as one of the new technologies driving the economy.

In Fig. 9.7 a sensor/actuator network implemented to get information from across the irrigation system described in Chap. 3 is depicted. In the diagram at the lower level, there are the farm sensors mainly sensing water level and flow but also temperature, humidity and radiation, in some cases a full local weather station can be incorporated. The pictures at this level are samples of these devices.

At the second level, all radio networked, there are the node and the repeaters concentrating the data gathered as well as preprocessing the information. In this way the network reliability is increased as redundancy, due to dynamic relationships between sensor data can be exploited. Again, the pictures at this level show some examples.

Finally, at the highest level, the nodes and repeaters are connected perhaps using an optical network with a central node, where all the information can be digested and used to manage the system.

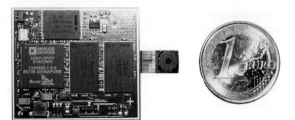

Fig. 9.6. Sensor dust compared to a Euro coin

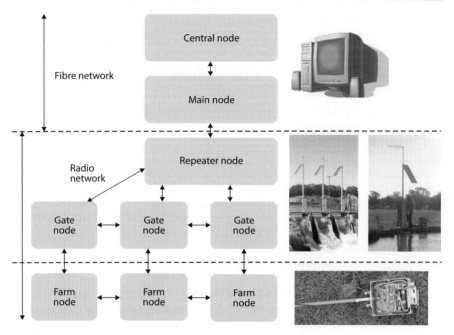

Fig. 9.7. Components of an irrigation sensing network

Sensors

Sensors witness what happens in the controlled plant. The sensing systems provide data about the process, as well as about the disturbances and their evolution over time. The information extracted from these data is the basis for all control actions.

9.3 The Controller

The role of the controller has been described before, here we discuss some of the simplest controller subsystems that are commonly in use.

9.3.1 Automata and PLC

The automatic washing machine, the car wash from Sect. 3.5 or a vending machine are examples of systems where the signals can be represented as binary valued and where a well-defined sequence of events determines how the process evolves. These systems are known as automata, and their controllers are also automata.

Their *complexity* stems from the fact that there are only a few operations {AND, OR, NOT, IF} and the signal value can only be either {TRUE, FALSE}. The ingredients are few, and hence even a simple (but sloppy) statement like *Lets wash the car* becomes a rather long (very precise) sentence in Boolean algebra. Programmable Logic Controllers (PLC) were invented to enable the systematic implementation of automata. Programming languages were developed to easily describe logic equations and rules.

> **Automata**
>
> An automaton deals with binary signals (On/Off) and events, triggered by binary signals (true/false).
>   Automata are modeled by logic statements.
>   Automata take actions using binary signals, based on decision that are logic statements.
>   Automata are finite state machines. The state has finitely many values, and there are finitely many transitions between states. They can be fully understood using Boolean algebra.

9.3.2 On-off Control

Relays are very simple components that can be used as the main ingredient in a simple control loop.

In the ideal case, the relay output, u, takes one of two possible values $\pm V$ depending on the sign of the input signal, e, as shown in Fig. 9.8a.

A relay combined with a threshold detector can be used as a simple thermostat: it switches the heating on/off depending on the room temperature being too low/too high with respect to the set-point. Clearly, on-off control is simple, and delivers a low quality of control. It is easy to appreciate that the controlled variable will be oscillating around the set point. The nature of these oscillations will determine the acceptability of this control.

Due to technical limitations, but also to avoid continuous on/off oscillations, a real relay has a slightly different characteristic. In Fig. 9.8b the presence of a dead zone is shown. The output of the relay vanishes as long as the input is inside the threshold. (This would work better in the thermostat, and the heater will last a lot longer.)

Relays may also exhibit some hysteresis. Figure 9.8c represents this behavior. The switching point depends on the whether the input signal increases or decreases through the switching region.

Relays are used in many applications, in particular for motor control, like in moving lifts, conveyor belts and cranes.

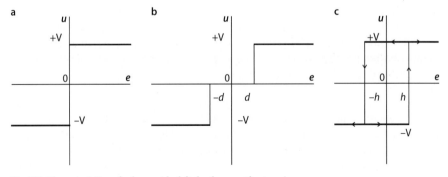

Fig. 9.8. Characteristics of relays: **a** ideal, **b** dead-zone, **c** hysteresis

9.3.3 Continuous Control: PID

The most commonly used local control subsystem in industrial processes is the so-called PID controller. PID stands for proportional, integral and derivative action.

P-Control

The simplest feedback is to make the input proportional to the detected error and such that it opposes the error: negative proportional feedback. This is the P in PID.

This proportional feedback is the basis for most simple controllers. There is an obvious draw back: no error, no control action. So if this controller works, and the plant requires an input to work as required, there will always be a residual error.

The greater the gain in the P-controller, the smaller the remaining error will be. Also, typically the larger the gain, the faster the system responds to reduce the error. Unfortunately as the gain is increased the system may become unstable.

Nevertheless, proportional controllers have been and are implemented with success in many applications. The Watt's governor (Fig. 8.1) is perhaps the best known P-controller. It was precisely the instability caused by such mechanisms in more advanced, faster engines that initiated the serious mathematical study of the stability of control.

PI-Control

To eliminate the drawback of the P-controller, an integrator (the I in PID) is added. This forms the so-called PI-controller. The control output is the sum of two terms: a proportional feedback and an integral of the error term.

In this manner, the error at the plant output can be zero, and there will still be a control action, or plant input as the integrator will retain an output, as it has accumulated, or integrated the past error.

There is an obvious question, which value does the integrator start with? This question has to be asked every time the process changes operational conditions. Typically, PI-controllers have so-called soft-handover routines that allow the resetting of the integrator in a manner that ensures that the controller is not responsible for nasty transients.

The design of such a controller is more complex, two gain variables have to be set, one for the proportional action and one for the integral action, and the integrator needs an initial value. Also, the integrator increases the dynamic complexity of the overall system. Hence the tuning of a PI controller requires more elaborate design, but the extra design freedom promises more performance in a larger variety of control situations.

PD

P-action is instantaneous. I-action remembers the past. Often some preview action is quite beneficial, and D-action provides this. The D stands for derivative, the derivative has the capacity to predict future values of the signal (within limits).

In particular in position control of mechanical devices, like antennae, the plant itself has high inertia and low damping. This results in relatively slow response and when resonances are present, oscillatory behavior that is unwanted. Damping can be provided through velocity feedback or indeed feedback of the derivative of position.

In the PD-controller the control action is thus proportional to the current error and is augmented with an action that is proportional to the current rate of change of the error.

The main concern in the PD-action is its sensitivity to noise. The derivative of noise is even more noisy creating spurious control action. To implement PD, noise must be filtered or suppressed.

PID

The PID-controller presents all three options. It is by far the single most popular control device with more than 90% of all simple (one output, one input) control loops in the world reportedly controlled using PID controllers.

> **Proportional Integral Derivative Control**
>
> The control action is the sum of three separate terms related to the current error, the integral of the past error, and a simple prediction of the future error.
>
> - **P** action is proportional to the observed error. It provides an immediate response to a current error.
> - **I** action integrates the past error to provide an enduring response. It aims to remove steady state error in a regulation environment.
> - **D** action anticipates the process response by taking action proportional to the derivative of the error. It aims to provide better damping, and faster response.
>
> The design involves the selection of four parameters, the gains for each of the actions, and an initial condition for the integrator.
>
> To improve on PID control, modern controllers mainly concentrate on a better future prediction and a better summary of the past, using more information about the plant model, disturbance model and noise properties.

9.4 Computer-Based Controllers

A microprocessor-based controller only deals with digital data. The basic control loop is shown in Fig. 9.9. The most common control algorithm, the PID, is often implemented using digital control. The main difference is that the signals into and from the controller are now sampled. The most common sampling technique is periodic, that is at equidistant points in time the controller receives inputs and delivers outputs.

During a sample period, the control output which is the input to the actuator remains constant. So the plant is effectively in open loop during the sample period. Provided the sample period is small, compared to the speed of the dynamics (think of the Nyquist sampling criterion) the distinction between sampled data control and analog control is small. Of course, the sample period must be large enough to accommodate the computations required for determining the control action. At present, only very demanding control situations require special purpose hardware design, as computing is rather cheap compared to sensors and actuators. The (resource) constraints in control are typically not generated by the control computer.

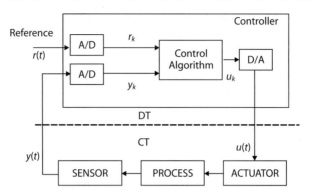

Fig. 9.9. Basic digital control loop

Nevertheless, digital control is not analog control, and computing resources are always finite. More importantly, computer algorithms are sequential in nature. So when many control loops start sharing the same computer hardware care must be taken to ensure that all control tasks are attended to on a fair basis. We mean fair from the system's performance point of view. Special operating systems, where time plays a direct role (so-called real time operating systems) have been developed to account for these difficulties.

The following are some of the issues that differentiate between the analog, real-time environment, and the sequential programming environment in the computer. The same issues come also to the fore when control data communication is happening over a shared communication network.

- There are delays associated with sensing, communication and computation. Delays affect system's performance.
- The sampling period which is the time between control updates may depend in general on the availability of the necessary resources. Sampling becomes an event rather than a periodic phenomenon.
- Signal timing is affected. Time stamping of all signals is essential, and requires network wide clock synchronization.
- Some measurements may be lost, or delayed.
- New failure modes, due to communication or computation errors appear.
- Verification and certification of control algorithm becomes much more complex due to the additional complexity in computing hardware.
- Local control becomes intertwined with other tasks like supervision, coordination, alarm monitoring, alarm responses. Resource allocation and resource guarantees have to be established to ensure performance.

In such a setting, the control laws are implemented in software. Often the control code is but a small portion of a very large code required to run the facility as required. The great advantage is of course the flexibility offered by the computer environment allowing us to conceive challenging control ideas. The disadvantage is that control fundamentally alters the behavior of the system, and as such it is often essential to guarantee its performance. The more complex the environment, the more difficult this task becomes. Nobody will accept an autopilot that may inform us as: *Sorry, the computer has momentarily lost the ability to steer, please wait till the system reboots.*

Discretization and Quantizing

The analog signals in the process must be sampled, to be treated digitally in the computer-based controller. See Sect. 4.4.2. Both the precision with which time and signal value can be represented impose limitations to how well control can be executed.

Codesign of Control and Its Implementation

Traditionally, the fields of control and computation have evolved without much interaction. Nevertheless, when dealing with control time is of the essence. Computer-based systems must react in real time: implementing outdated control actions is at best useless, and at its worst disastrous.

The current trend is therefore to look simultaneously at the control requirements and the availability of computing and communication resources, in order to determine what can be achieved. See also the CCC discussion in 8.7. Where systems are mission critical, codesign is automatic, as for example is commonly practiced in flight control design.

> **The Brain of the Controlled System**
> The computer-based controller determines the input to the process.
> The computer differs in many ways from the human brain. The computer lacks creativity, reasoning, liveliness, character, but in a control environment some of these weaknesses are important benefits as they imply repeatability, improve reliability, and guarantee verifiability.

9.5 Actuators

Actuators are transducers converting a low power signal coming from the controller into a high power signal capable of manipulating the process.

Typical actuators are valves to manipulate material flows, motors to move loads and amplifiers to provide electrical voltage and electrical power.

Ideally the signal into the plant should be the control signal as it was intended by the controller.

Smoothing

At the very least the actuator must convert the digital number from the control computer into the physical domain of the input of the plant. Invariably this conversion involves some form of interpolation as the control command is quantized, and comes at discrete points in time, whereas the signal to the process must be analog and smooth. A zero-order-hold interpolation, keeping the control command constant over the sample period is presented (see Fig. 9.10). This is a discontinuous signal and presents a problem as actuators cannot respond instantaneously, since they always have inertia (they

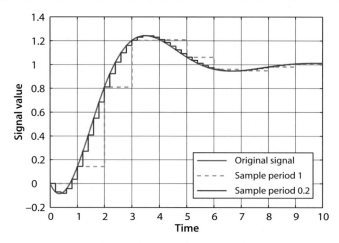

Fig. 9.10. A continuous time signal sampled at two different sample rates

store energy). So the actual input signal into the plant will react to the discontinuous signal from the controller, and will never equal it. It simply tries to follow it with some smooth response. In applications where this is critical, the model of the process dynamics should account for the actuator dynamics.

Nonlinearities

Actuators are never linear devices. Other than smooth nonlinearities like slight changes in the gain, actuators always have a limited range of operation.

Dealing with a valve the output is bounded between being fully open, providing maximum flow, or fully closed. The valve in Fig. 9.11, presents its internal operation. The vertical displacement of the valve axis, x, determines the flow f through the valve body.

Nonlinearities like thresholds or dead-zones and hysteresis are common in actuators.

An electrical motor does not start to turn until a minimum voltage has been applied, due to the presence of static friction.

Typical in mechanical actuators is backlash, which can be modeled as hysteresis. This is due to the gap between gear teeth (see Fig. 3.12).

Disturbances

Being power amplifiers, the actuator is always dependent on an external power supply, and its characteristics.

The flow through a valve not only depends on the valve opening but in the differential input/output pressure of the fluid. If the pressure is changed, the valve gain is changed.

The power amplifier gain is not at all constant.

In order to avoid these problems, typically actuators are always placed in a dedicated analog feedback to linearize the response. As long as the actuator gain is sufficiently large this is not a problem. The situation is identical to what happens in operational amplifiers, see Sect. 7.3.

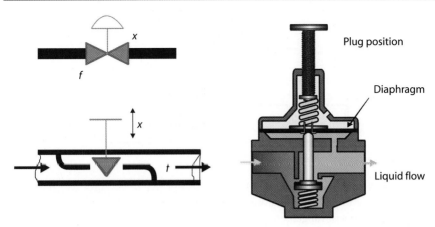

Fig. 9.11. Symbolic and schematic representations of a valve

Process Coupling

For complex processes, there are many actuators that manipulate different process variables. Invariably, these actions are interrelated (through the process) and will interact in the process to generate the response. The process model, and the controller must account for this interaction, otherwise erroneous responses will result.

Multivariable control is precisely designed to deal with this situation.

Discrete Actuators

The D/A converter can be avoided if the actuator admits a digital input. On/off actuators, such as switches, fall into this category. More complex are stepper motors which take a pulse train as input and generate appropriate discrete shaft rotations in response.

9.5.1 Smart Actuators

Modern actuators are really actuator systems, with locally embedded micro-processors for control, and signal conditioning and the ability to communicate across a variety of different control networks. They provide for alarm conditioning, preventive maintenance and calibration. Often they can be reprogrammed via the control network.

Similar to sensor dust, there is currently a lot of research in developing actuator and transducers that can be integrated in silicon in systems-on-chip technology.

9.5.2 Dual Actuators

Often it is difficult to combine both the required dynamic range as well as the required precision in the same actuator, see for example the pH regulation problem touched on in Sect. 8.8.1.

Dual actuators in which one is able to address the range, and another one is able to address the precision are a way forward. An example is the dual actuator used in hard

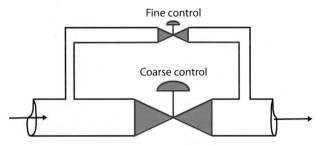

Fig. 9.12. Dual actuator: coarse and fine control

disk drives. A voice coil motor provides for a fast but coarse response, whilst a piezo-electric actuator provides for an even faster response with very high precision but over a very limited spatial range.

This situation is typical when you want to manipulate a flow in a wide range but with great precision as well. In this case, see Fig. 9.12, a large valve is used for coarse control across the entire dynamic range, and a parallel much smaller control valve provides for precision.

> **Actuators**
>
> The actuators are the brawn of the controlled system.
>
> Actuators are typically considered as part of the process, because they involve powered components subject to uncertainty and disturbances.
>
> Modern actuator are systems in their own right, with local control, communication and signal conditioning capabilities.

9.6 Comments and Further Reading

For texts devoted to measurement principles and transducers, we refer the interested reader to Bannister and Whitehead (1991), Klaassen and Gee (1996) and in particular Sydenham (1984) where control is the driver for measurement. Unfortunately, in as far as these books deal with technology they are quickly outdated. A book that deals more with the philosophy of measurement is Berka (1983) (after all measurement goes to the heart of the scientific method).

Similarly, the type and nature of actuators depend very much on the application area, and is mainly driven by technology. A fairly recent overview is Janocha (2004). Most books tend to focus on a single industry or single actuator technology. The process control industry is serviced by the massive *instrument handbook* which gets regularly updated (Lipták 1995).

Programmable logic control is old technology, but has widespread use. A modern reference is Rohner (1996). The modern PLCs are much more than just logic controllers, often they have evolved to become fully fledged industrial computers suited for generic real time control applications. Standardisation in this field is rather limited. Some of the main suppliers are OMRON^{TM}, Siemens^{TM}, Allen-Bradley^{TM} (a division of Rockwell Automation^{TM}), ABB^{TM}, Schneider Electric^{TM} and GE Electric^{TM}. A text devoted to one commercial family of PLCs is Dropka (1995), see also also Matic (2003) for a text on a different family of PLCs.

Soft sensors, and in particular Kalman filters are treated in (too) many books. The number of patents dealing with Kalman filters in one way or another is a true testimony to its remarkable success in many diverse applications. A www-site devoted to the Kalman filter is, for instance, *http://cs.unc.edu/~welch/kalman/*. Here you can find the seminal paper by Kalman (1960) as well as entire lists of Kalman filter related books, from the beginner level to expert user and research monographs. An introduction from a mathematical point of view is Catlin (1989). The Kalman filter is also used in time series analysis, and financial engineering (Wells 1996).

Considering control, communication and computation (CCC) as an integrated design problem is attracting a lot of research at present (Graham and Kumar 2003). An older study anticipating much of what is presently common technology is CTA (1982). Smart dust, embedded system technology and zigbee (Gislason 2008) are some of the technologies driving this research activity. Flight control is another area where sensing and actuation are always considered together with control design. A good introduction to practical flight control systems is Pratt (2000).

The PID controller is the most used and also abused control algorithm in the world. It is ubiquitous in the petrochemical, process and manufacturing industries. A book dealing exclusively with PID design is Astrom and Hagglund (2005). In Johnson et al. (2005) the authors provide an overview of PID technology as well. The variants and different physical incarnations of PID controllers are almost uncountable.

Chapter 10 Control Design

10.1 Introduction and Motivation
10.2 Control Design
10.3 Local Control
10.4 Adaptation and Learning
10.5 Supervision
10.6 Optimization
10.7 General Remarks
10.8 Comments and Further Reading

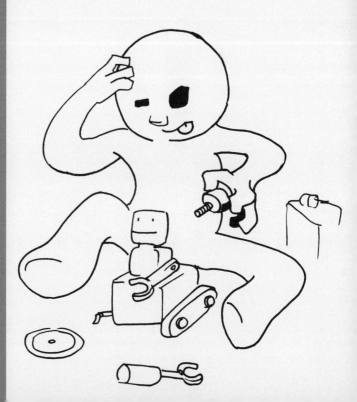

Chapter 10

Control Design

> *You know what you want (assume that)*
> *You know what you have (you must!)*
> *but ... do you know what to do?*

10.1 Introduction and Motivation

The natural environment together with its processes is already designed. We merely observe, analyze and use it. Using it we typically alter the environment, but only when we modify the environment *by design*, do we call it an engineered system.

The design for an engineered system typically starts with characterizing the service the system must provide together with the expected performance level at which this service must be delivered. In control design we start with specifying what we expect the system to do. Typically this is done by describing features of the outputs of the system, and describing the limitations of what inputs can be used. To summarize our objectives we often define a control performance measure. This is a quantitative index that enables the evaluation and hence comparison of various alternative implementations of the controlled system. It makes selection of a preferred implementation easy.

In analysis and design, two main approaches are used: top-down or bottom-up. Both are incarnations of a divide and conquer approach for problem solving. Often both are used in an iterative manner.

In the top-down approach we first look at the whole system, through the observation of some key (external) signals. How can we influence its behavior? How does it operate, how does it respond to our external inputs? What does this achieve?

As W. Gibbs stated, the whole is always simpler than the sum of its parts[1]. Nevertheless, there is clearly no point in joining subsystems to build a larger system if the latter does not do more than the sum of its parts.

The overall system point of view may suffice to understand the key aspects of the behavior. If this is sufficient for our purpose, no further analysis or design is necessary. This will also be the case if we want to use the system under consideration as a building block in a larger system.

More often than not, we are not satisfied with the behavior and want more. In this case we need to zoom in, breaking the system up in subsystems so as to identify more signals that is more degrees of freedom to work with. At this level of greater detail we have then indeed the potential to do more and expect more. As usual, this comes at a price. As we start to unravel the internal operation of the system and its structure, the analysis and design task becomes more complex. Each subsystem, identified through

---

[1] Everyone that attempts to simulate a complex system from the bottom up point of view should take note of this.

a number of signals is now treated as a system in its own right and our analysis of the behavior is repeated at this scale. For the design process, we need now to understand the subsystems' requirements as well as the overall system requirements.

Ideally to come to a thorough understanding of the subsystems, we want to isolate these from the influence of other subsystems, and the environment. In the synthesis step, the interactions between the subsystems will be crucial in determining the overall behavior of the system. If we are working with a physical realization of the system, it may not be feasible to achieve isolation of subsystems, without destroying the overall system.

The top-down approach can be iterated till we reach elementary subsystems whose behaviors are easily understood in their entirety. This level signifies the end for analysis. Design typically stops at a level of sub-subsystem detail where modifying subsystem behaviors is no longer realistic or feasible.

To some extent, we followed this approach in presenting the applications in Chap. 3.

The analysis of a system's behavior is however not complete until we also complete a bottom-up reconstruction of the overall system from the subsystem building blocks so to understand how the interconnections of subsystems really brought about the overall system's behavior. This step verifies that we did the right thing (or not).

Top-down is also very powerful in design, in particular when we have a very good understanding of the type of building blocks we are going to use. First we specify the overall objectives or service requirements. Then we unpack these in lower level specifications of sub-behaviors that when combined achieve the higher level specifications. We then design a logical structure of subsystems according to these behaviors, which when interconnected achieve the required system goals.

In turn, or in parallel (and that reveals the power of top-down), we then design each subsystem in isolation with the aim of delivering the required behavior. This procedure is repeated for each subsystem, until we arrive at an elementary or well-known system for which we either have a realization or have a guaranteed recipe for its design.

Finally we assemble each subsystem, verify the specifications at this level, and work our way to the top, until we are satisfied that the overall objectives have been met. Sometimes, we will need to retrace our steps when a test fails at a certain level. It may unfortunately mean that we have to retrace all the way down to the lowest level to identify the real issue.

The advantage of the top-down design method is that

- it allows readily the re-use of known systems,
- it accelerates the design process by creating natural parallel pathways, allowing design teams to work independently towards their own design goals
- it enables a natural test and evaluation hierarchy, making debugging more systematic.

Top-down design is commonly used when there is a lot of experience in design of similar devices, services or systems.

In designing completely novel functionality, or a non-existent device, the bottom-up approach is more common. Starting from basic subsystems with known behavior, novel and perhaps challenging interconnections are envisaged to deliver the new functionality. Assembling these subsystems we build up a new system. This has to be tested and its behavior thoroughly established. The process is iterated till all design objectives are satisfactorily met, and safe operation can be guaranteed.

Most design methods use a combination of both top-down and bottom-up methods, depending on the stage of the design, available design experience, and the encountered difficulties that have to be overcome.

Rather than experimentation with prototype systems, computational experimentation using computer-based models for the systems plays an increasingly important role in (control) design.

Control design, which is particularly concerned with the design of a so-called control subsystem, follows along the same general principles. Though rather than seeing control design as the design of a control sub-system which has to interact with a plant or process, it is far better to see control design as an integral part of the overall system design. Unfortunately more often than not, typically control design is practiced after the plant has been designed. When control is indeed a retrofit measure this is unavoidably the case, and one has to live with it. When it can be avoided, it is normally a big mistake to see control design as following system design. Indeed a control subsystem can only operate within the behavior of the plant, and shape this behavior. Hence by virtue of the plant behavior, desirable (controlled system) functionality may either already be lost, or can be too difficult to realize no matter what control subsystem functionality we envisage. Like in all design, there are fundamental limitations, things control cannot do. Revealing these limitations is one of the main objectives of control theory. Co-design of plant and controller subsystems, seeing them both as integrated parts of the overall system is always to be preferred. Unfortunately there are very few texts dealing with this.

As presented in Chap. 8, control uses two main interconnecting structures:

- *open-loop or feed-forward* action to impose inputs on the process, using knowledge of the plant's behavior and possibly from external disturbances or reference signals;
- *feedback*, which relies on information from the process itself to determine appropriate actions;

as well as their combination in order to achieve the controlled system objectives.

A complex system with many interconnected subsystems, controllers and systems should be hierarchically organized. A local controller uses local information derived from the subsystem (and perhaps from of a few of its neighbors) to act on a subsystem to deliver local specifications and to minimize unwanted interactions between subsystems. A coordinating higher level controller informs the local controllers of their precise objectives so that the overall system is orchestrated to achieve its global objectives.

Depending on the envisaged autonomy the control subsystem(s) must cater to such diverse objectives as described in Sect. 8.3. Modern design, based on models, makes heavy use of optimization tools, to drive incremental improvement in system performance, hence the importance of the performance measure. Usually the available control design methods and their tools are specialized to a particular control objective. The pursuit of multiple objectives may be achieved through the subsequent exploitation of the remaining design freedom, as no single design objective should really exhaust all design freedom. If at any stage there is indeed no design freedom left, the closed system performance may not be very robust, and difficulties are to be expected. In such circumstances it is better to revisit the design specifications. Alternatively, a

Chapter 10 · Control Design

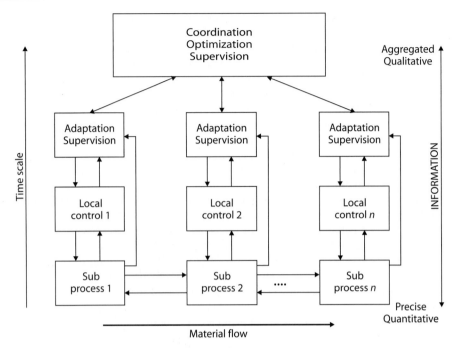

Fig. 10.1. Control levels in an integrated framework

multi-criteria optimization methodology may be pursued. Such brute force approaches are typically hard or too hard. Certainly as system complexity grows, the notion of optimality is rather elusive, and one often settles for incremental improvements.

A normal divide and conquer approach is to arrange the objectives in a hierarchical manner as depicted in Fig. 10.1. Time scale and/or spatial scale separation ideas should be exploited to reduce the complexity, i.e. by addressing different objectives at different levels in the control hierarchy. Altogether, these stages constitute a holistic approach to control design.

At the lowest level, close to the fastest physical sub-processes, measurements from the process are used to inform local control actions. Information flows fast, is precise and quantitatively accurate. This gives rise to tight control.

This information, probably filtered and aggregated with other local information is transferred to a higher level control where decisions about adaptation, optimization and/or supervision are made. This can happen on a much slower time scale without upsetting the local controllers, and allowing for a gradual transition towards better overall behavior.

Finally, at the upper levels, the gathered information is aggregated and mainly qualitatively expressed to coordinate the plant operation based on the partial results from the subprocesses.

In this chapter control design issues at these various different levels are briefly discussed. The reader interested in control design will be referred to a variety of references, most of them including computer-based control design aids and worked examples. Ideas will be illustrated using applications including those we introduced in Chap. 3.

10.2 Control Design

Dealing with a well-defined system or subsystem, the following steps (somewhat stylized, as design is typically rather messier) are common in control design.

First we specify the control objective(s), this really means we understand what can be done with the plant[2], and that we have a plant model suitable for this control design purpose[3]. Realistically though, it is very likely we do not really know how to specify what we want, nor know what can be done. A safe place to start is to repeat what has been done before, and then explore how and what can be done better. Design specifications that work in the first iteration are most likely too easy. Also, a best design does not exist, and normally it is enough to have an incremental improvement on what has been done before.

A. Model

- Specify the control objective(s).
- Identify the available inputs, domain and range. At the very least the available inputs must be able to affect the control objective in an obvious manner. If this is not the case, a plant redesign or a control objective redesign or a combination of both is called for.
- Identify the available outputs. Ensure that all performance objectives can be accounted for through measurement. If not, question the validity of the control objective, or invest in further sensors (either direct sensors, or soft or virtual sensors).
- Identify disturbances. Are they measurable? What model do we have?
- Identify reference signals. Is preview information available?
- Acquire a (control) model for the process using all identified signals. Quantify the reliability of the model. What are its limitations?

The dimensioning of actuators, their dynamic range and capability should be considered at this point in time, and if necessary (when their dynamic response is similar to that of the plant) these must be incorporated as an integral part of the (extended) plant. Likewise the sensors, their dynamic range, and response as well as accuracy have to be matched to the task at hand. Also, if sensors have a significant dynamic response this has to be included in the (extended) plant dynamics before control design is commenced.

At this moment, the process, possibly augmented with actuators and sensors has been provisionally defined. A block diagram as in Fig. 9.2a, using all the corresponding variables should be now available. A description of uncertainty, both from a model as well as signal perspective should complete the model specifications.

We are now in a position to continue with the control subsystem.

---

[2] There is no point in asking control to let water flow up-hill when the only energy source is derived from gravity.
[3] Typically this means a mathematical model, but a model fit for this purpose, not necessarily a model that represents the plant behavior completely.

B. Analysis and Design

- Choose a suitable control structure (feed-forward, feedback, combinations, nesting, decentralized) and draft the control block as in Fig. 9.3. The chosen information graph, which measurements link with which inputs is critically important. Experience plays an important role here (verifying the sensitivity of the plant outputs with respect to its inputs, provides a good starting point for pairing signals). The simplistic, meaning all measurements are available for all inputs may provide the greatest flexibility but is also the most complex and for large scale systems typically too complex. Unfortunately there are no simple ways to select information structures. Moreover their enumeration grows exponentially with the number of inputs and outputs, which all but eliminates the possibility of considering all possible structures and selecting the most suitable one.
- Transform the control objective(s) to a number of control objectives appropriate for the selected control structure. Given the structure specify what each control subsystem has to pursue, so that the combined effort delivers the overall goal.
- For each sub-controller and their associated objectives select an appropriate design methodology to deliver the sub-controller dynamics.
- Analyze sub-system performance, and verify if design specifications are met.

At this point, both the model and the control algorithm are all defined. In the last phase, we can now evaluate and test, to see if the choices we have made deliver as per the design expectations.

C. Implement and Validate

- Validate the design first by simulation then experimentally using hardware-in-the-loop technology. Tune and re-tune the controller parameters as required. If this step fails, verify the specifications and go back to analysis and design, typically this reveals that some of the models or assumptions were inadequate. This step is normally performed bottom-up.
- Define a controller implementation. In the case of digital controllers select the hardware and software to fulfill the control requirements. There are a number of commercial software packages available that can translate simulation software designs into micro-controllers and software programmable hardware solutions to facilitate this step. Validate the implementation against simulation and then experimentally, again in a typically bottom-up manner.
- Synthesis: implement and commission the control, actuators and sensors with the plant.
- Test and evaluate the controlled system performance across a wide range of realistic operational conditions through the entire operating regime.
- Test and evaluate the controlled system under a variety of expected failure conditions. When safety considerations are critical these are to be an integral part of the design specifications and take obviously precedence over performance.
- Finalize commissioning, hand over the controlled system to the end user, provide for continued monitoring of the successfully controlled system.

Control Design

In summary, to design a control system we must:

A. *Know the process to be controlled from a control point of view.* What can and cannot be achieved? *Define what is to be expected after control*, and ensure that this is within the (acceptable) behavior of the process (which includes the actuators as well as sensors).
B. Select a control structure and design a control system.
   Compare different control structures and define the final controller.
C. Fine tune the controller parameters.
   Implement, test and validate the solution.

10.3 Local Control

The steps in the design process outlined above, labeled A and C are common to all control design tasks. The actual control design steps B may be qualitatively different and are elaborated in the sequel.

10.3.1 Logic and Event-Based Control

In this case the signals of interest are mere binary, representing logic True or logic False. Typically control problems in this category are about scheduling the plant, either in open loop or in closed loop.

Simple examples are a vending machine or a car wash like the one described in Sect. 3.5 and revisited in Sect. 8.4 and Sect. 9.3.1. In such applications it is the sequence of actions that matters. To illustrate the design procedure consider a process of filling bottles, similar to that shown in Fig. 10.2.

There are a number of conveyor belts that transport a row of bottles or palettes. Dealing with separate bottles, each palette arrives at a particular location, where a preprogrammed operation is performed. At the start, the bottle is picked from a feeder palette and placed on the belt. It is cleaned, dried, filled, closed, labeled and finally packaged and assembled into a transport box.

All the activities are predefined, clearly delineated, and typically no problems are expected. Each operating station is fully programmed and executes its task as soon as a bottle is in position, and completes this task in the prescribed time that the belt remains

Fig. 10.2. An automatic system to fill bottles

stationary. From a global point of view, the activity at each station is based on a feedforward command. Locally, closed loop control and analog signals handling may be used to complete the prescribed task. For example, the bottle filling can be either performed in open-loop using a time-based approach or may be based on a closed loop using the filling level to determine that the bottle is full. In principle the overall system is hybrid, requiring both discrete as well as continuous variables. Nevertheless the progress of the system, from a global point of view can be captured by purely logic-based signals: a bottle is/is-not in place; is/is-not filled, and so on. The actions are equally binary: the belt moves/stops, the bottle is open/closed, the bottle is full/empty. Any failure in completing a task raises an alarm, which is also a logic signal, task complete/task not complete.

The signals are binary valued and Boolean algebra (dealing with ones and zeros) can be applied. Basic logic operations such as AND, OR, NOT, and their compositions can be used to formally describe the system dynamics. Moreover logic can provide a formal design specification and verification. Computer tools exist to synthesize a design and even deliver the computer program that will implement the entire controller.

The controller recipe or program will itself be a string of logic statements. For instance, the bottle filling station, may have a specification of the form:

```
IF "bott-in"(=1) AND "bott-empty"(=1) THEN "start-filling"(=1)
IF "bott-full"(=1) AND "filling-removed"(=1) THEN "bott-move"(=1)
```

There are many techniques for dealing with the analysis and design of logic based controllers. They play an important role in computer science. Originally such algorithms were implemented in Programmable Logic Controllers (PLC)s. In Fig. 10.3 a state machine of a sequential system is depicted, and it is composed of:

- *The enumerable states*, S1, S2, ... Sn. These represent the possible working conditions of the system. They are like the working stations in the bottle filling system. At each state the system performs a given task or a set of tasks or a sequence of set tasks until completion is reached, or another condition takes precedence.
- *Transitions*, t1, t2, ... There is a finite list of all possible logical conditions in the system environment. If the process is in one state enabling a certain transition, and if this transition then happens (becomes True), the automaton evolves to another state as prescribed by that enabled condition. Both the transition itself must come true or be enabled and the process must be in an enabling state before the transition can take effect.

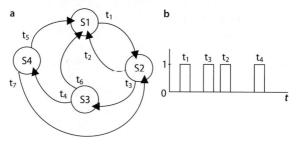

Fig. 10.3. Sequential system: **a** state machine; **b** transitions timing

Conditions, enable state transitions, and are represented by the a finite set of feasible transitions.

Let us explain the behavior of the automaton in Fig. 10.3a. State S1 enables transition t1, whereas, for instance S3 enables transitions t4, t5 and t6. The time line of the enabled transitions is shown in Fig. 10.3b. The automaton is initially in state S1. It will move to S2, as soon as t1 is enabled. The next transition that can occur is t3, and the system will move to S3. At this point in time t2 is enabled but nothing happens because the system state does is not enabled for this transition. The system will remain in S3 until t4 becomes true.

Observe that if t2 would be true before t3, the automaton's evolution is different. In general, transitions occur asynchronously, they can happen at any (analog) time, and are solely governed by the conditions that logically enable them to occur.

The design problem consists in determining a set of states and transitions so that the automaton evolves from any possible initial state to the desired state. Unwanted states have to be avoided. Preferably the design is such that the automaton can recover if for some reason or another a transition or mishap disturbed the state away from the normal evolution.

For systems as simple as those in the examples considered so far, the design appears to be a rather easy task, and it is. The difficulty lies in the complexity of the system. It is not difficult to find automation problems in the manufacturing industry where the automaton has thousands of binary conditions and transitions. The number of states of such systems grows geometrically fast. By way of illustration n binary conditions leads to 2^n possible states. That is, for $n = 10, 100, 1\,000$ this leads to a number of possible states such as $1\,024$, 1 followed by 30 zeroes, and 2^{1000} respectively. The latest is a number larger than the number of atoms in the universe. In such cases, simple graphical representations, like the one used in Fig. 10.3a are not useful. Special tools and appropriate software is required to deal with such systems.

The theory for modeling, control synthesis, and verification of automata is well-developed. There are numerous computer tools for their implementation. The enormous complexity these design tools can deal with underscores some of the great advances made in control technology.

10.3.2 Tracking and Regulation

The most common local control objective is the regulation of a process variable to a prescribed value or a small range of values regardless of disturbances or changes in the plant dynamics. Tracking a reference signal with prescribed precision is also very common.

In Chap. 8 a number of possible local control structures have been reviewed, including feedback, feedforward and combinations thereof. There is a very extensive literature dealing with these types of problems.

Usually in this context signals are analog, but may either be represented in discrete time, as sampled signals or in continuous time. Disturbances may be described using statistical or stochastic methods, or deterministic or even a mixture. Time and frequency domain approaches as well as linear and nonlinear techniques have been developed. Systematic as well as heuristic methods, optimization-based or recipe-based designs are available. Without being exhaustive control design synthesis depends on such aspects as:

- **The process or plant model.** Most modern control design methods are model based and exploit some form of optimization method or use formal verification ideas. Invariably they require a mathematical description for the plant and the disturbances. Plant models can be:
  - single input/output or multiple input/multiple output.
  - linear or nonlinear;
  - time invariant or time variant;
  - input-output or state-space-based;
  - represented in continuous time, or discrete time, or use event-based timing, or indeed a combination of time axes;
  - stochastic or deterministic;
  - precisely described, or be a member from a (large) set of possible systems that needs to be treated simultaneously;
  - have a finite state space description or require an infinite dimensional state (the state itself is a function) space description.

  All possible combinations can be contemplated.

- **External signal models.** Similarly disturbance models, or reference signal models are varied, and must suit the particular problem at hand.
  - Disturbance/reference signal models can be linear or nonlinear.
  - Disturbances/references can be deterministic, a member of a large class of signals;
  - Disturbances/references can be stochastic, with certain properties like a sample mean being prescribed by the model;
  - Disturbance signals can influence the system at the input, output or a particular internal signal, or a combination of all;
  - Disturbance signals can modulate the system in an additive or multiplicative, or more generally nonlinear way;
  - Disturbance characteristics may be fixed in time, or time varying.
  - Disturbances/reference signals may be measurable or not, and we may have preview information or not;

- **Heuristics.** Some control design is model-free. In this case, the design is based on heuristics, known to work for a large class of systems. Experience plays an important role in such approaches.
  - Recipe-based design rules. For example, the famous Ziegler-Nichols PID tuning process is such a method[4, 5]

---

[4] John Ziegler, 1909–1999, American chemical engineer (1933, Washington University) and instrument designer famous for his work on PID tuning which he developed together with Nichols at Taylor Instruments in 1940.

[5] Nathaniel Nichols, 1914–1997, American control engineer, educated at the University of Michigan, and famous for his work on the PID tuning rule, and control design more generally. The International Federation of Automatic Control (IFAC) society has since 1996 an award and medal in his honor to recognize design progress in the control field.

- Rule-based designs, capturing the behavior of an operator, or expert user of a system. Often used in the automation of relatively safe processes (like taking a digital photograph or making a video recording).
- **The control goal.** Depending on how the control objectives are being expressed:
  - as an index to be optimized, with or without constraints on the signals (which may be expressed at points in time, over time intervals, or terminal constraints);
  - in the frequency domain (bandwidth, maximum gain, gain and phase margin)
  - in the time domain (time response, oscillations, delays, overshoots, undershoots, range conditions on signals, gain conditions on signals, operator ease conditions);
  - as a set condition using a combination of frequency and time domain requirements.
- **The operational regime.** Control design can deal with normal operation or fault mitigation, uncertain conditions, changeable modes and switching conditions. These different operating conditions typically also require different models for the plant.

The Ziegler-Nichols PID Tuning Rule

In a unity feedback loop, with a PID controller to determine the plant input, conduct the following control experiment. Set the control gains for integral and derivative action to zero and increase the proportional gain until the plant output exhibits a sustained oscillation. Denote this gain as Kc. Measure the period of oscillation and denote this as Tc.

- **P-control.** Set the proportional gain to $0.5\,Kc$.
- **PI-control.** Set the proportional gain to $0.45\,Kc$ and set the integral action gain to $0.55\,Kc/Tc$.
- **PID-control.** Set the proportional gain to $0.6\,Kc$, set the integral action gain to $1.2\,Kc/Tc$ and the derivative action gain to $0.08\,Kc\,Tc$.

Consider a process with a two-degree of freedom controller structure as depicted in Fig. 10.4 (see also Sect. 8.6.1). Assume that all signals are scalar valued. For convenience assume that all blocks are representing linear operators or linear systems.

The blocks have the following meaning. H is the measurement device operator, also used to suppress some measurement errors, here represented by the disturbance n. F performs reference signal filtering to smooth out possible sudden changes in the reference signal r, which in any case could not be tracked by the plant. K_{ff} and K_{fb} are respectively the feedforward and feedback controllers. Disturbance signals d and n enter at an internal part of the process and its output respectively. The plant operator is $G_2 G_1$, here split in two blocks from left to right first G_1 followed by G_2. The disturbance d is added to the output of G_1. The disturbance n is added to the output of G_2 to form the output y that is measured, and used for feedback.

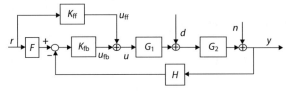

Fig. 10.4. 2DoF control diagram

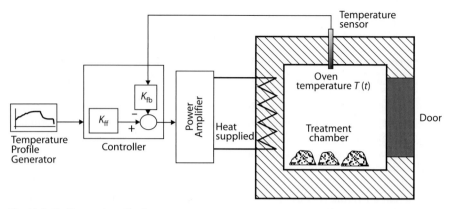

Fig. 10.5. 2DoF control application

The feedback controller K_{fb} is designed to provide stability in the loop and disturbance rejection properties. The feed forward controller K_{ff} is to provide good tracking response. It augments the performance already achieved by the feedback control loop. The feed forward controller must itself be stable, and does not play a role in determining the stability of the loop, nor can it.

This structure is useful in dealing with a heat treatment control problem as depicted in Fig. 10.5. The objective is to apply a predefined temperature profile r to whatever is inside the oven. Here stability is of lesser concern (but still important, as we could not tolerate oscillations in the oven temperature). The feedback controller K_{fb} is primarily designed to keep the desired temperature in spite of disturbances such as opening of the door or changes in the load (which are reflected through the disturbance d).

Filtering the measurement noise inherent in the sensing device is performed by block H. It should have the effect that $Hn \approx 0$, yet the output of the plant should be measured accurately. There is always a trade-off.

The tracking controller K_{ff} is designed to ensure that the plant output tracks the temperature profile r over time.

As in this oven treatment situation, even in the general case (Fig. 10.4), the feedback controller K_{fb} must achieve disturbance rejection. It does so by reducing the size of the operator linking the disturbance d to the output y wherever d is large. This operator[6] is:

$$y_d = M_d d \quad ; \quad M_d = \frac{G_2}{1 + G_2 G_1 K_{fb} H}$$

Note that K_{ff} does not play a role. The objective of $\|M_d\| \ll 1$ can be achieved by making K_{fb} large. It is not possible to make K_{fb} arbitrarily large, as stability must be preserved. Typically K_{fb} will contain an integrator, as this will eliminate the effect of a constant disturbance d altogether.

The tracking controller K_{ff} can now be tuned for tracking. Its objective is to ensure that $y \approx r$. The tracking operator is

---

[6] It can be computed using the block-diagram algebra introduced in Sect. 5.6.

$$y_t = M_t r \quad ; \quad M_t = \frac{G_2 G_1 K_{fb} H}{1 + G_2 G_1 K_{fb} H}\left(F + \frac{K_{ff}}{K_{fb}}\right)$$

Disturbance rejection and tracking are not independent. But in general, we can first design for disturbance rejection, and stability and then address the tracking design by selecting K_{fb} to obtain $M_t \approx 1$.

Some possible feedback design strategies are:

- *Tuning controller parameters* Given a controller structure, tune its parameters according to a procedure, like a rule of thumb, or parameter optimization routine with the aim of finding an acceptable setting. This works well in most simple applications, or where extensive experience exist, like when a general solution (say PID controller) is already known to be able to do the task at hand, and the only remaining task to reflect the present situation is to select the actual parameters.
- *Assign closed-loop dynamics* Determine K_{fb} to get appropriate dynamics of the closed-loop. There is a variety of options. The most popular choice is to use state feedback to assign a particular closed loop response. The procedure often involves optimization.
- *Improving time response* Augment the plant with a controller, and address the full time response characteristic. Typically this is a large scale optimization problem that nevertheless can be solved efficiently for linear systems.
- *Shaping the frequency response* of the loop, $G_2 G_1 K_{fb} H$, similar to the time domain, but rather working in the frequency domain, which often provides for more insight, but often leads to more cumbersome optimization problems.
- If the disturbances are measurable, add extra *feedforward* action such as $u_{ff} \approx -1/G_2 \cdot d$ in order to approximately cancel the foreseen effect of the disturbance.
- *Optimizing* a given index (this is the hard bit, what is a good index?) to achieve the *best* (that is with respect to the index) disturbance rejection, the best robustness against process model uncertainties, or the best tracking performances.

10.3.3 Interactions

So far we have dealt with examples where there is a single input controlling a plant. Most processes are more complex. Typically there are multiple inputs, effecting multiple outputs. Moreover, in such situations we almost always have competing objectives.

Clearly for every output signal that needs to be regulated or tracked, we need an input, otherwise the problem is ill-posed in general. It would be really easy if each input only affected the one output (in which case we have multiple examples of single input single output systems) but life is not like that.

A classical approach (building on what we know about single loop situations) is to pair inputs and outputs (as in u_1 is used to control y_1, or reacts on deviations in y_1 from its reference signal r_1, and u_2 is used to control y_2 and so on). The pairing is typically performed to reflect the dominant effect, i.e. u_1 mainly affects y_1 and much less the other outputs, and so on. Because u_2 does affect also y_1, from the point of view of designing a control law for u_1, the u_2 input (and all the others as well) will be treated as a disturbance. Of course all inputs are measurable, so that we can use a feedforward controller to assist in rejecting the effect of the other inputs. This is called *decoupling*.

Unfortunately, pairing with decoupling does not always suffice, and design must be considered as one multiple input multiple output synthesis problem.

Pairing (with decoupling) clearly simplifies the design problem considerably as we are only interested in designing $2m$ (or $2m + m(m-1)$ if we use decoupling) controllers when there are m pairs. The full multiple input multiple output design would consider the design of $2\,m^2$ control laws. The additional freedom of multiple input multiple output design, augurs well for achieving better performance but it is of course more complex. Moreover, in case of faults (like measurement stations failing) it may be much more difficult to reconfigure the complex solution, than the simpler one. Considering all possible operational conditions, it may well be that a more constraint, controller with lesser performance is to be preferred above the high performing complex controller. Selecting an appropriate control structure is a difficult task in general.

Consider by way of example the boiler control discussed before (see, for instance, Fig. 7.8). In principle, an obvious choice for pairing seems to be: control the water level by the water flow and the boiler temperature or pressure by the fuel flow, as shown in that Figure. Nevertheless there is a strong internal interaction because any time the desired temperature is adjusted by the fuel flow, the steam/water equilibrium changes and the amount of water required in the boiler is also to be changed. So whenever the fuel flow is affected, the water flow should also respond. Equally, adding cold water to the boiler, requires extra heat input. It seems plausible to design an integrated controller that takes care of these obvious links to ensure that the water flow and fuel flow work together to maintain the water level and temperature as desired. In this case, experience indicates that the pairing approach using a 2 degree of freedom in each loop does not do too badly.

A schematic for a paper-making machine is presented in Fig. 10.6. The main properties that determine the paper quality are the specific weight and the moisture content (thickness also plays a role). The drum speed determines the throughput in the plant, and this is of course of great economic value. There are several options to control these variables, but they are clearly not independent. The manipulated variables are typically the pulp flow, at the exit of the headbox, the steam flow inside the drying cylinders, the drum pressures, the sheet speed, the sheet tension, and so on. A good pairing is paper-weight coupled to pulp-flow and paper-moisture coupled to steam-flow. In this case, the rest of the input variables that regulate the throughput are considered as disturbances and they need to be decoupled.

The state-of-the-art for multiple input multiple output system design is well-advanced, in particular for linear systems. Automated synthesis packages that can produce reasonably acceptable controllers from specifications and plant models are available.

> **Multi-Input-Multi-Output (MIMO) Systems**
>
> To independently control a number of process variables, at least the same number of manipulated variables is required.
>
> Usually, these variables are internally coupled. MIMO control is based on either the principle of
>
> - *decoupling*, pairing inputs with outputs, and treating other inputs as disturbances to be rejected, or
> - *cooperation*, or integrated, if all the output variables are linked into the control law for all the inputs.

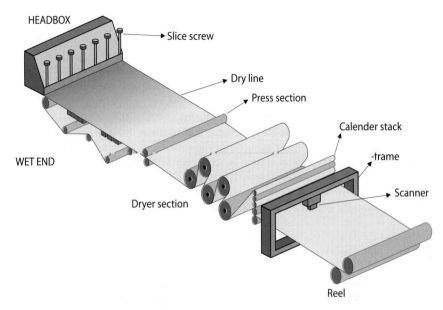

Fig. 10.6. Control options in a paper mill

10.4 Adaptation and Learning

Much of control can be successfully completed using models that are linear and time invariant systems. This is the case even when the process itself is not linear and not time invariant. Indeed control by its very nature is in the first instance pursuing the regulation of signals. As long as signals are close to their reference signals, linear models typically are sufficient enough to capture the essence of the process behavior.

Nevertheless, at times, linearity is too restrictive an assumption, or the plant characteristics change with time, or the reference signals vary so dramatically over time that the true nonlinear and/or time varying nature of the plant cannot be ignored.

Nonlinearity is not a useful system characterization, it encompasses everything, including linearity. It is sometimes, and rightly so, referred to as the absence of definition. It comes then as no surprise that no control design method can deal with nonlinearity in its full generality.

Adaptive control is a design methodology developed to cope with systems that exhibit natural time scales, either as a consequence of some time varying or nonlinear behavior in the process (Astrom and Wittenmark 1988; Anderson et al. 1986a). Essentially the assumption is that at any point in time, and for a short term the process may be adequately modeled as linear, even time invariant, but over extended time scales these local-in-time models may drift far apart.

Adaptive control has its roots in the early 1950s, when people attempted to use classical frequency domain techniques to control an experimental aircraft with highly nonlinear control characteristics.

10.4.1 MRAS: MIT Rule

Assume we want our controlled process to behave similarly to a given model. The model reference adaptive system (MRAS) schema (Landau 1979) uses the difference (error) between the model and process outputs to tune some controller parameters to reduce this difference. The controller parameters and the adjusting mechanisms can be simple or sophisticated.

The so-called MIT rule (Mareels and Polderman 1994) was first proposed to automatically compensate for a single parameter that captured the essence of time variation (perhaps due to nonlinear aspects) in the process. It is represented in Fig. 10.7. The single parameter is compensated by using an adjustable or tunable control parameter. The tuning of the parameter is based on how much the process output deviates from the desired output (which assumes that the rest of the model is well-known), and the parameter is tuned so as to minimize the integral of the error squared.

$$\frac{d\theta}{dt} = -k.e.\frac{de}{d\theta} \tag{10.1}$$

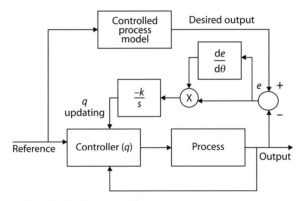

Fig. 10.7. A simple MIT rule adaptive control law

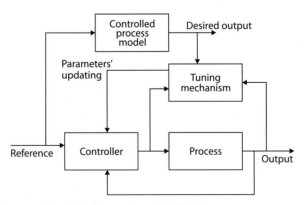

Fig. 10.8. Adaptive control: model reference adaptation, MRAS

This control law is however deceptively simple. Its dynamics and the dynamics of the controlled system are rich and even today only partially understood (despite a fairly extensive literature on the topic). As long as the assumption of time scale separation is respected, its behavior is quite acceptable and intuitive.

Nowadays, adaptive control deals with far greater complexity than envisaged by the MIT rule, which gave rise to its development. The approach proposed by Landau generalizes the model reference approach, comparing information coming from the output of the process and that from a closed loop reference model. The adaptation mechanism, as shown in Fig. 10.8, computes the updated value of the controller parameters based on the evaluation of the error in the controlled process output.

10.4.2 Self Tuning

In the self-tuning approach, pioneered by Aström, the goal is to adapt the controller parameters to perform as close as possible to the control requirements based on the direct or indirect knowledge of the process model, as shown in Fig. 10.9.

In the indirect case (Fig. 10.9a), the identification block estimates the parameters of the process which are used to recompute the controller parameters. Based on input/output process information taken at a faster rate, the parameter updating is carried out more slowly to avoid the strong coupling between the two closed loops. Similarly, in the case of direct adaptation (Fig. 10.9b), the estimation block directly estimates the (slowly varying) controller parameters from the input/output information derived from the plant.

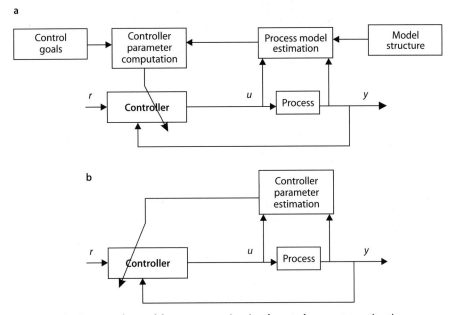

Fig. 10.9. Adaptive control: **a** model parameters estimation; **b** control parameters estimation

10.4.3 Gain Scheduling

Another approach, simpler but quite efficient in many practical applications, uses a selection of pre-computed controllers each based on a different process model adapted for a particular operational condition of the process. During normal operation, the most appropriate control action is selected according to the actual operating conditions. This approach goes by the name of gain scheduling. It is suitable when all operating conditions are well-known in advance. Typically a collection of representative operating conditions is chosen and a corresponding set of controllers is com-

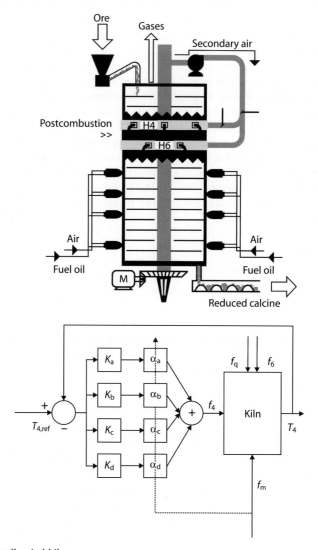

Fig. 10.10. Metallurgical kiln

puted off-line. The control design can be of any type and the controllers may even have different structures, adapted to the operational needs. To work well in practice the switching between control modes should not be too frequent and must be smoothed so as not to disturb the input signal. For example, flight controllers are typically designed on this basis. The scheduling is performed based on a combination of speed and height measurements for the aircraft. At the transition between controllers, the control action is typically computed as a smooth interpolation between the previous and the newly desired input signals.

The idea is illustrated with the control of a metallurgical kiln, as represented in Fig. 10.10. The objective of control in this reduction furnace is to optimize nickel recovery while minimizing fuel consumption and environmental contamination. This demands precision control of temperature and gas composition in the furnace. Controlling the temperature of a multiple hearth furnace is a difficult task (Ramirez and Albertos 2008). Fast and extensive changes in operating conditions occur, complicated by non-linear and time-varying behavior of the process and interaction between the different variables. Temperature in level 5 is strongly dependent on the temperature in the adjacent levels 4 and 6.

Four operating conditions are selected based on furnace loading. For each a linear model that relates temperature (the output) to injected air flow (the input) is obtained. The responses under the different operating conditions vary greatly and no single controller can properly work across all operating conditions. For each of the selected operating conditions a particular PID controller is designed. Under normal operation of the furnace, the current loading of the furnace is measured. The current loading is then represented as a weighted sum of the operating conditions used for design, and the actual air flow is then obtained as the same weighted sum of the control actions computed by the PID controllers associated to these operating conditions. This is shown in the bottom of Fig. 10.10.

10.4.4 Learning Systems

The dream plug-and-play controller would learn what to do based on past experience. In full generality such plug-and-play controllers are simply elusive, but within a sufficiently constrained environment, learning control is quite feasible. Learning is an advanced form of adaptation, and adaptive control ideas are at the core of most learning control systems.

A learning controller could have the structure depicted in Fig. 10.11.

There is a set of possible controllers denoted as the option set which typically contains a safety controller (for when all else fails). The information from the plant is processed by the current controller which determines the current input actions for the plant; but also by the adaptation block, which adapts the controller parameters to maintain performance; the evaluator block, which computes the adequacy of the current controller; and the clustering block, which identifies different modes of operation. Based on the outputs from these blocks, the decision system determines which is the best controller to be used and the supervision system implements the transfer between the current controller and the newly selected controller.

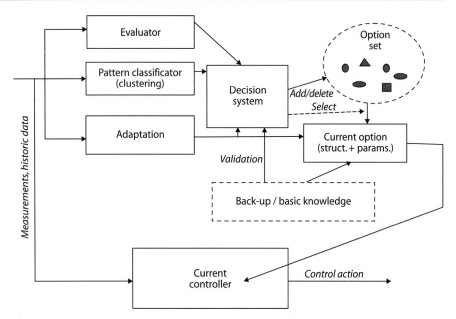

Fig. 10.11. Plug-and-play control: starting from scratch

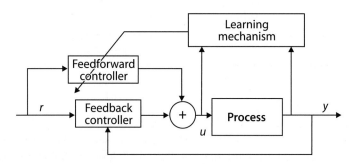

Fig. 10.12. Repetitive learning control

Learning control is well-suited to the control of repetitive operations. This is common in many robotic applications. For instance, a welding robot must follow a prescribed trajectory to weld parts in a car chassis. The action is repeated for every chassis. Similarly pick and place robots collect parts and place them in a particular order, over and over again. As the operation is repeated, it becomes possible to learn from the past cycle to improve the next cycle. From cycle to cycle the control action can be improved, i.e. the robot learns how to execute the desired operation. This is shown in Fig. 10.12. This control strategy goes by the name of *repetitive learning control*.

The actual control action will consist of two parts: an open-loop or feed forward action which is refined through the learning action, combined with a feedback action which is necessary to cope with unavoidable disturbances and sudden process variations that cannot be learned. This is illustrated in Fig. 10.12.

10.5 Supervision

At the highest level in the control hierarchy, a supervisory control system coordinates all other controllers and may even be able to adapt the information structure. A supervisory control uses alarm signals to reconfigure the control subsystems so that the overall performance gracefully degrades in the presence of faults. In case there is sufficient redundancy it may even be possible to be fault-tolerant: the performance is not degraded despite the fact that certain faults have occurred. In case the reliability or safety of the process and its users is challenged, the supervisory controller will initiate a fail safe procedure, or perhaps a full shut down.

For example in the irrigation system (see Sect. 3.3), when a significant rain event occurs in a particular region, the affected canals go into shut down mode, and all feeder canals must stop delivering water into the affected canals. The new control objective is to store as much water as feasible in the canals and all excess water must be passed safely through the system, bypassing all farms. All set points for water levels and flows are reset for extreme conditions. As previously discussed, some adaptive control implementations require a supervisory level to decide the best controller among a set of previously computed ones. In general though, supervisory control deals with switching between operational modes, shutting down or starting up, initiating emergency procedures and so on.

Supervision can be implemented in a number of different ways, using human in the loop or human supervision, assisted by diverse levels of automation and automated reasoning, or fully automated:

- *Alarm-based*. For each alarm a particular system response sequence is determined in advance. Depending on the alarm the response may affect a small sub-system or the entire system. Typically the aim is to restore pre-alarm conditions, or a new safe operational condition, or in extreme conditions may lead to a shut down sequence. An example is the exploitation of the electricity distribution grid. The supervisory system implements the sequence, and adjusts the sub-control systems accordingly.
- *Performance-based*. Based on information gathered from the process, the supervision algorithm computes a number of scenarios, or compares a number of precomputed scenarios and selects the next actions and adjusts the local sub-controllers to address the new situation. This is typically feasible only within the normal operational conditions of the system. This approach can be combined with the alarm-based approach.
- *Model-based* supervision is the most advanced solution for (adaptive) control. Using on-line data the model for the plant is updated, as well as the control objectives and a global optimization algorithm monitors and tracks the entire process. This requires a great investment in a realistic model, and typically requires large computer resources, and the highest levels of monitoring and automation, which have to be justified against the benefits of operational performance. A space shuttle, a space laboratory, aircraft control and critical processes such as complex scientific instrumentations (colliders, particle accelerators) fall into this category.
- *Goal-based*. The supervisor switches the sub-controllers on the basis of the relevant goal and the present operating conditions. This can be based on a table look-up, or simulation basis. This scheme is depicted in Fig. 10.13. As shown in this figure, the supervisor may decide on a new control structure.

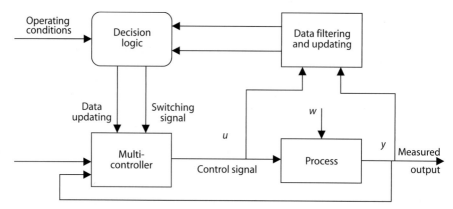

Fig. 10.13. Goal driven supervision

Most of these supervision principles can be used in combination, but eventually at some level will be subject to human supervision. Truly autonomous systems that can interact with their environment as well as with humans are still an elusive research pursuit.

10.6 Optimization

In general optimality means little. In principle, every controller is optimal, with respect to some appropriately selected criterion. There are some interesting theorems to this effect in the control literature. A trivial way to appreciate this, is to generate a problem that only has the one solution, this is then the unique and optimal solution.

Nevertheless optimal control theory plays an important role in control design. Meaningful problem formulations pay attention to:

- *The plant model.* A clear, computationally efficient model of the plant fit for the purpose of control is needed.
- *The constraints.* Signals in the model cannot be arbitrary. The constraints are an integral part of the problem definition. A model for the disturbances and reference signals is equally required.
- *Design goal*, which is the criterion with which to compare the different feasible solutions and to decide which one or which collection is/are best. Time optimality, energy efficiency, maximal suppression of disturbances, whilst guaranteeing stability are typical design objectives. Design objectives always involve inputs and outputs of the model[7]. The design goal may incorporate constraints (signals cannot be too large, must not oscillate and so on). In optimal control the model is always a constraint, as the model shows how inputs and outputs are dynamically related.
- *Options/inputs.* What are the variables and/or parameters that may be selected? What are the constraints they have to satisfy?

---

[7] If the criterion does not contain output signals from the model, the model is irrelevant. If only output behavior is penalized, the problem is typically ill-posed, mainly leading to unrealizable behavior.

There are many approaches to design optimal controllers. The literature on optimal control is substantial, and strongly linked to calculus of variation in mathematics. In general, optimal control design requires numerical techniques and many problems quickly turn out to be hard, especially if they involve discrete and analog signals at the same time. Even defining feasible solutions, let alone searching for an optimal solution may be difficult. On the other hand, for the very respectable class of linear system models with criteria quadratic in the input and output signals, systematic solution methods are well-established. Moreover, they provide a lot of insight as to why optimal designs are useful, and in what sense they provide *robust* solutions.

10.6.1 Controlling the Read/Write Head of a Hard Disk Drive

As an example of optimal control let us illustrate the position control of a magnetic hard disk read/write head, as depicted in Fig. 10.14. Thin structures of steel, known as suspensions, supporting the read-write heads are attached to the end of the actuator arm. The actuators are typically voice coil motors. There are two control goals corresponding to two modes of operation: track seek and track follow. In track seek the actuator moves the head from one track location to another, and this must be accomplished in minimum time, without moving the head out of bounds. In track follow mode, the read and/or write operation is executed, which requires the head to follow as precisely as possible the spinning track, which due to imperfections in the mechanical drive mechanism (for example eccentricity) does move relative to the head. The track follow mode is a disturbance rejection control problem.

More modern assemblies for read/write heads use dual actuators. A low inertia piezoelectric transducer/actuator is mounted on the slower voice coil motor. This piezoelectric transducer is very fast, but has only a very small reach, perhaps only the distance between two tracks. The dual actuator combination allows one to achieve both speed and accuracy. The response to a command to change the reading track is represented in Fig. 10.15. In the graph, the contributions from the fast and slow actuators are split to illustrate the different time scale of their actions. The graph clearly shows that the dual actuator response is a significant improvement on the conventional single actuator.

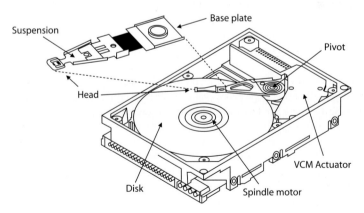

Fig. 10.14. Disk drive

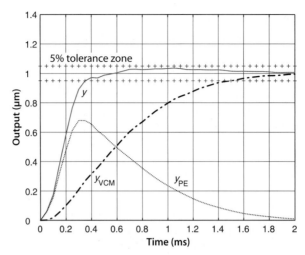

Fig. 10.15. Transfer between tracks

10.6.2 Model Predictive Control

A well-accepted optimal design approach developed first in the petroleum industry is the so-called Model Predictive Control. It refers to a class of feedback algorithms that compute the inputs based on a receding finite horizon optimal control problem, exploiting realistic models and constraints. The basic idea can be expressed in the following way:

MPC definition. Given a plant model (say $y = Gu$), and a desired reference trajectory for the controlled output (say y_r):

1. Denote the present time as t.
2. Compute the input u over the interval from present time t to an instant $t + T$ in the future so that it minimizes some criterion involving, the output error $y - y_r$ and the input u while satisfying all constraints.
3. Apply this optimal input u over the time interval from t to $t + h$, where $h \ll T$.
4. Measure y over the interval t to $t + h$.
5. Set $t + h$ as the present time t. Go to step 1.

The optimal control strategy computed in the first step is an open loop input strategy. The advantage is that this step can be evaluated even in complex situations and satisfying hard constraints. There are few methods that can deal with constraints so effectively. In the next step this control is implemented for a short period of time, and then re-evaluated. In this way, the open loop strategy is moderated by the actual measurements, and becomes effectively a closed loop control as the input is responsive to what actually happened in the system under control. It is a most effective control strategy. Many industrial applications of MPC have been reported (Camacho and Bordons 1995). MPC enjoys strong robustness properties as well and in general provides a good strategy for control (as long as a reasonable model is available to compute the input in the first step).

Advanced Control Structures

The first goal of a control system is to make the process to operate appropriately within its normal operational envelope. Open-loop control or a simple feedback control loop are the first options.
Based on the availability of more resources like

- computing power,
- sensors and data acquisition systems,
- more and better actuators,
- more knowledge about the process and its disturbances,

and when the controlled behavior requirements are demanding, then one may want to consider, in order of complexity

- optimal, robust, controllers,
- adaptive, complex nonlinear controllers,
- hybrid, switching-based controllers,
- hierarchical, artificial intelligence based controllers.

Keep in mind that in control, reliability and safety are often more important than performance. The golden rule in design is: *keep it simple*, and remember *anything that can go wrong will go wrong*.

10.7 General Remarks

When approaching a process control design problem, it is not uncommon to concentrate first on elementary sub-systems. The temptation is great, as the most complete, optimal and beautiful designs can be worked out at this level (as the problem is easier, more well-posed and so on). Nevertheless it is the performance of the whole system that matters. Any optimization at the subsystem level may well be totally wasted, because quite generally the coordination of many locally so-called optimal systems does not yield the optimal system behavior.

The global system matters. Failure in one subsystem may be disastrous for the whole. The first aim has to be to achieve a feasible, reliable design for the entire system. Fine tuning and optimization can follow later. In any case, optimization must be seen from the overall system performance and not from a local subsystem's point of view. Even then, global optimization may not be the aim, as safety and reliability must always take precedence.

For example in the irrigation canal system, global optimization demands that all water levels are used to determine all gate positions in the canal. This multiple input multiple output design is feasible on paper, but will never be implemented because it relies on communication that is too fragile to be guaranteed. A suboptimal, decentralized design using pairing and decoupling works well, and has much higher safety and reliability features. It is such a solution that is commercially implemented in Total Channel Control^{TM}.

Also, as most control design is model-based it pays to remember that all models have always limited validity. In control design it is therefore important to ensure that the obtained performance does not rely on model properties beyond their region of validity. For instance, given a model of a motor it is easy to conclude that the higher the input voltage is, the shorter the time for the motor to reach its desired speed. The problem is that the voltage that can be applied is limited and beyond this limit the motor burns. More often than not, the model's domain of validity is however poorly understood, requiring validation and testing of any control design before it can be fully accepted.

10.8 Comments and Further Reading

An excellent introduction to feedback systems and design (with more mathematics and rigor than we dared to include here) is Astrom and Murray (2008). This book is also available on the world-wide-web. In general, most books dealing with control do require a command of a university level introduction to calculus and linear algebra. Domain oriented control references are not very common (Albertos et al. 1997).

As to the importance of automata the foundational work by Turing and Von Neumann cannot be overlooked. In this context see Turing (1992) and Von Neumann (1958). Some of the designs created by Von Neumann, Turing and Norbert Wiener for a truly autonomous machine are still very much dreams being pursued in systems engineering, and computer science alike.

Discrete event systems, a generalization of automata used in the modeling and design of modern manufacturing systems are discussed in Cassandras and Lafortune (2008). Modeling is the main aim of David and Alla (2005), which provides a very didactic introduction. Synthesis for discrete event systems can be found in such texts as Zhou and DiCesare (1993).

Multivariable control design is discussed and developed in a number of texts on control design. To name but a few Albertos and Sala (2004), Goodwin et al. (2001), Green and Limebeer (1995), Boyd and Barratt (1991). These are not accessible without a working knowledge of university calculus, including complex functions, and elementary linear algebra.

The ideas about self-learning, adaptive and/or self-tuning control, are quite old, going back to A. Feldbaum's formulation of the dual control principle in 1960. The principle exposes the conflict between learning and regulation. Regulation reduces complexity so that there is nothing to learn, and learning needs signals that excite the dynamics, and so contravene regulation. Also, because any learning automatically leads to non-linear control, understanding the behavior of adaptive control is hard. It is now understood that adaptive control in general implies some form of chaotic behavior. Texts dealing with adaptive control ideas are Mareels and Polderman (1994) and Goodwin and Sin (1984). In Anderson et al. (1986a) an effective adaptive control strategy based on time scale separation ideas is presented.

Frequency domain ideas are introduced in a classical manner in Doyle et al. (1992), and a more modern and more algebraic treatment of the frequency domain ideas is found in Green and Limebeer (1995). For a historical account and the influence of the foundational work by Bode and Nyquist, see Mindell (2004).

Ideas of dual actuator control and how this applies to hard disk servos, an example we have used a few times are explained in Chen et al. (2006).

Model predictive control turns optimization into a control methodology. The ideas and application areas are well-presented in Maciejowski (2002). The methodology was first developed in the petrochemical industry, and its theoretical development followed the success in industrial applications (Camacho and Bordons 1995). The control methodology is presented in some detail in the Instrument Engineers' Handbook (Lipták 1995), where it is quoted as the most effective control approach to deal with large scale, nonlinear processes.

Chapter 11: Control Benefits

11.1 Introduction
11.2 Medical Applications
11.3 Industrial Applications
11.4 Societal Risks
11.5 Comments and Further Reading

Chapter 11

Control Benefits

> *Life itself, the ecology of our planet
> and the wealth of nations all depend on
> intricate, precariously balanced, and
> poorly understood feedback loops.*
>
> Anonymous

11.1   Introduction

Despite the fact that control is integral to how the engineered world as well as the natural world function, it is hardly ever clearly identified. Control is truly a hidden technology. Indeed control and feedback consist in the intelligent design of control algorithms, and are implemented through the subtlety of interconnecting subsystems. The following analogy may illustrate how (well) control is hidden: control relates to systems, as the mind relates to the brain.

In the engineered world, control and feedback are not restricted to industrial, manufacturing or agricultural applications. They are equally central to the service industry. Feedback plays an important role in risk management. Control ideas are used to optimize the flow of passengers through an airport. System theoretic ideas have been used in such environments as managing negotiations. In our economy, feedback is at the core of the behavior of supply and demand, and the reason why central banks use interest rates to stimulate the economy. Feedback loops are at work at all levels of processes, from the room temperature controlled by a thermostat to the behavior of the world climate.

The greatest benefits of feedback are no less than the very existence of life as well as the stability of our climate. No economic or engineering benefits could possibly rival this accolade. Moreover, it is equally true that to a large extent our present day standard of living as well as our longevity could not possibly exist without the ubiquitous application of control and feedback. Indeed from the very beginning of the industrial revolution, feedback has underpinned the relentless growth of our economy.

At some level our engineered control systems may be seen as automating and performing activities that human operators could perform. In such situations, control realizes benefits through improved reliability, repeatability and consistency of the performed activities. An example in this vein is a cruise control in a car, allowing the driver to use their attention span for more important things other than speed control. As an additional side benefit, the fuel economy of the car is improved. Far more influential though are the development of systems or behaviors that are not achievable without the use of control or feedback: satellite flight, the internet, an automated irrigation system enabling water markets, a robot exploring the Martian surface, a catheter guided inside the circulatory system without the need for surgery, or an automated wheelchair that responds to thought patterns of a quadriplegic person and so on.

In this chapter we mention some of the benefits of control and feedback that also serve as motivation for their continued further development and application.

11.2 Medical Applications

The impact of control in biomedical engineering and more generally health applications is enormous.

The two most elementary yet critical engineering contributions to longevity are the availability of clean water for sanitation and the refrigerator, because it removed salt from our diet. Neither running water, nor the refrigerator can exist without feedback and control.

Sophisticated control systems improve the quality of life for people affected by heart or other basic organ malfunction. For example, hemodialysis apparata depend on adaptive control technology (pressure control, anticoagulation, and so on) to provide their essential service. In a hospital's emergency department, the monitoring of vital parameters of patients combined with appropriate reactions (from both nurses and doctors as well as machines under their supervision) improve patient care immensely.

In surgery, the development of miniaturized robotic devices using wireless communication enables doctors to carry out surgery or monitor body functions with minimally invasive techniques.

Figure 11.1 shows an ALF-X system with a physician positioned at the operator console in the control room. This is a tele-robotics surgical system with haptic feedback jointly developed by SOFAR S.p.A. and the Joint Research Centre of the European Commission[1]. It is a modular system that may consist of up to 5 manipulator arms remotely commanded through one or two surgeon's consoles, having the capability of remotely delivering 1:1 scale tactile sensing, which is in practice the transmission of the forces exerted by the tip of the surgical instrument to the handle of a haptic device manoeuvred by a surgeon. This feature enhances the surgeon's perception with respect to manual procedures. Together with the accurate control of the forces applied along the instrument shaft, tactile sensing potentially improves the efficiency and safety of the intervention. The system can be used for any type of laparoscopic intervention easily adapting to operative conditions and changes, as patient's orientation and close placement of access ports, without increasing the intervention time.

Fig. 11.1. Laparoscopic surgical intervention

---

[1] http://ec.europa.eu/dgs/jrc/downloads/jrc_tp2770_force_estimation_for_minimal_invasive_robotic.pdf.

The doctor is entirely oblivious to the complexity of the required control of surgical instrument so as to realize the desired opertion. Both feedback and feed forward control techniques (using sophisticated models of the manipulators) are essential to guarantee the success of the intervention.

People with physical disabilities suffer much, as they become dependent on others in their environment for the simplest of tasks. The emotional consequences (for themselves and their immediate support circle) can be devastating. The economic impact of disability, due to both the loss of participation in economic activity by the affected individual and the cost to society to provide a meaningful life, is enormous. Automated control systems provide for much progress in this area. A powerful example is the development of the bionic ear: a feed forward control device that restores near normal hearing ability to the profoundly deaf by directly stimulating the auditory nerve.

There are many other medical applications that rely on control and feedback. A small collection may serve to illustrate the tremendous impact: heart pacemakers respond to physical activity, insulin pumps respond to glucose levels in the blood, motorized prosthetic limbs, muscle controlled or even EEG controlled prosthetic devices, voice controlled wheel chairs, electronic nerve stimulators used in some treatments of depression and epilepsy. The emerging field of bionics is in its infancy but shows enormous promise.

Domotics

Automation in the home, domotics[2], heavily relies on feedback.

Most of our consumer goods (not just the toilet) use feedback to enhance or indeed define functionality. Refrigerators, freezers, ovens, central heating and airconditioning all use simple temperature feedback loops to deliver what we want. Washing machines, driers and dish washers use at the very least an automaton, but many have rule-based logic to minimize water usage and improve energy efficiency (some with an internet connection can be remotely monitored and receive software updates and preventive maintenance through the manufacturer's website). Some dishwashers, washing machines, and driers use active sound feedback to suppress motor noise to make them quiet. Most likely your video camera, or still camera uses auto-focus, image stabilization and auto-white balancing, all feedback techniques to allow you to focus on taking the shot. Your head phones (and your mobile phone) suppress ambient sound by feedback to allow you to enjoy your favorite music as if in a quiet environment, despite all the noises in the environment. They use automatic gain control to ensure minimal distortion, and some even cleverly adapt the sound field to create impressions as if you were in a cathedral. Ever wondered why speakers that are featherlight can produce incredibly nice sound, whereas they weighed a ton in the past? Feedback and feedforward are used to pre-distort the signal sent to the speaker so as to compensate for its transfer function, and thus neutralize the poor intrinsic speaker quality.

The above is simply the present, there is an entire revolution of domotics about to enter our daily lives. Lights, heating, security, and audio functions all reacting to our

---

[2] Domotics, a fusion of domus (Latin for home) and informatics.

presence, and possibly remotely controlled through an internet connection may seem plausible enough. What about a self vacuuming home, or a robot that does the vacuuming (including getting rid of the collected dust)? Self cleaning windows? A robot for mowing the lawn (not a sheep) that knows to respect your flower beds?

11.3 Industrial Applications

Control engineering is as much a part of the industrial revolution as the steam engine itself. The productivity gain achieved through moving away from the cottage industry to mass production with guaranteed product quality underpins much of the economic success and rising living standards in the last century. None of it could have been implemented without sophisticated control embedded in the overall manufacturing processes.

Control is essential to make complex systems consisting of many interacting subsystems work well. For example such large scale engineered systems, as a telecommunication network, the electrical power grid or a water, gas or oil distribution network can only function well because of control. In the electrical distribution network supply and demand are constantly matched by regulating the frequency of the mains (either 50 or 60 Hz). The quality of electricity service demands the appropriate regulation of the voltages in the network. The internet relies fundamentally on feedback control for its stability and functionality by throttling the supply of information packets from the sources (computers and phones) so as to not exceed capacity on the communication lines. All chemical, food processing plants, indeed all large scale manufacturing plants rely in a fundamental way on control to deliver consistent product quality. A modern car could not possibly be assembled without tight control of all the dimensions of the mechanical parts, manufactured across a wide variety of different automotive part suppliers. A typical computer chip, manufactured with nano meter precision, could not function without embedded control in its circuitry that compensates for the imperfections of the manufacturing process.

Control and feedback are used to overcome the effects of disturbances, and the inherent variability and uncertainty that occurs in all processing operations and raw material properties.

In the engineered (as well as the natural) world control delivers **benefits** in terms of

- *Complexity*. Allowing one to build more complex systems as a collection of interacting subsystems by ensuring that each subsystem behaves in a prescribed manner.
- *Safety*. Systems remain within operational limits, suffer less wear and tear and require less maintenance. Monitoring of system behavior essential for feedback enables preventive maintenance.
- *Efficiency*. Systems are controlled to use less primary resources (energy, raw materials), and produce less waste products. Learning and adaptive control can fine tune behavior to optimize for efficiency.
- *Functionality*. Systems behave more appropriately as unwanted behavior is suppressed through feedback and control, or even more spectacular, unnatural or rare behavior may become enabled and generic through feedback and control.
- *Quality*. Control produces more repeatable, and more stringently defined outcomes in the face of unavoidable uncertainty and disturbances.

All of these benefits are often translated into economic benefits or productivity gain terms, so as to enable the justification of the investment in control and feedback. How long is the pay back term for the envisaged investment in control? Pay back can either be realized in terms of additional income generated per unit of cost, and/or a reduction of costs to realize the same income. In the process industry, control is often implemented to realize savings in operating costs, but also to satisfy more stringent operational conditions, like minimal carbon footprint, or minimal environmental impact. These can be achieved through a reduction in the use of raw materials, energy, by improving the yield of the process, or by reducing the amount of waste materials and pollution. Control and automation can be justified on the basis of increased plant throughput by reducing the operational margins, without a need to invest in additional plant capacity. Equally important is the use of control and feedback to improve product quality and product uniformity, so as to command a better price in the market. Control can also be justified by improving the agility of the manufacturing process, and reducing its response time to changes in operation or product requirements.

We will elaborate on a few aspects.

11.3.1 Safety and Reliability

Typically the primary aim of a control system is to perform a control activity, that is to generate the appropriate input signal to a given system so that the output satisfies design requirements. As we have seen any control involves sensors to monitor the output, to verify that the system performs appropriately, as well as actuators needed to act on the plant. Typically these sensors and actuators allow for more than just the implementation of the immediate control objectives under normal plant operations, and additional activities such as providing for safe(r) operations are invariably pursued:

- *Learning.* Monitoring the controlled system behavior enables the user to learn more about the system and over time improve not only the control but also the safety record of the system.
- *Safety and maintenance.* Sensors can provide the information on which to base alarms, with which to identify or even predict a failure mode or emergency so that preventive maintenance can be implemented.
- *Fail safe.* Actuators provide also the capacity to act under faulty conditions and to avoid major disasters through a fail safe intervention strategy. Of course actuators provide points of failure as well, as do sensors and the control computer itself. Redundancy of actuators and sensors, as well as the computer servers that interpret the data and compute the control actions, is the normal operational strategy to safeguard against these failures in critical plant equipment.
- *Adaptive behavior.* In some systems the control and information structure (which sensors and actuators are active, and how does the information flow from sensors to actuators, and which computer algorithm is used) can be changed in response to overall system requirements. Such ultimate flexibility is often difficult to exploit, invariably poses challenging design problems (as the complexity of design increases dramatically with the number of different information structures), but may be used as an advantage in obtaining better operational conditions.

It is often the case that the actual computer code that implements the normal control operations of the plant is only a minuscule part of the entire computer code which governs all modes of operation, including foreseeable emergencies and failure modes. This is even the case in biology, where the part of the DNA that governs normal behavior is only a minor fraction of the entire DNA, as indeed most of the DNA is devoted to safeguarding functionality from reproduction errors.

There are only a few applications where safety is so critically important that it becomes the major driver for control. Aircraft auto-pilot flight control and the operation of a nuclear power station are in this category. Notice that in both instances, it is still a legal requirement to work with a human in the loop.

11.3.2 Energy, Material or Economic Efficiency

The purpose of control systems is typically to meet certain output specifications keeping the system variables within the normal operational range. More often than not there is sufficient freedom left in the process under these conditions to also pursue efficiency from either an energy consumption or raw materials use point of view. If not, one may consider and pursue a trade-off between efficiency on the one hand and performance specifications on the other. For example the cost benefit in the reduction of wasted energy and/or materials may justify a reduced income due to a lesser quality manufactured product with a higher yield of acceptable product.

In the production of steel in a hot and cold rolling mill (as the one depicted in Fig. 11.2), one may appreciate the tremendous amount of energy required to get to the final uniformly thin steel sheet. Any improvement in the control of this process typically leads to energy savings, and reduction of wasted product. Similarly, most industrial processes benefit from control through energy savings, reduction in the usage of raw materials, and reduction in waste products.

Fig. 11.2. Steel rolling mill

The irrigation example (Sect. 3.3) provides another illustration. The main driver for automation is to deliver water on-demand where and when required by the farmer, as this maximizes on-farm productivity. The other driver is to deliver water without spillage, so as to achieve maximum water efficiency. Where the cost of water is sufficiently high, the latter will take precedence over on-demand water delivery. Indeed based on purely economic rationalism, demand should be scheduled in function of the water price so as to optimize overall productivity or economic return. It is this economically moderated demand schedule that the control should implement. Moreover, without the control and its associated information infrastructure, this objective is entirely elusive. Nevertheless the social ramifications of this suggested water distribution strategy based purely on economic considerations may be rather non-trivial and require a substantial and appropriate policy framework as well.

11.3.3 Sustainability

At present the quest for a more sustainable operation, using less energy, or less raw materials, and producing less waste is a major driver for innovation in industrial systems. What matters is not only that the system performs as designed under nominal conditions but that its operation is such that it has a minimal environmental footprint.[3]

It is clear to us that control will be instrumental to achieve a more sustainable way of working. Of course this requires investment: instrumentation, analysis and design.

11.3.4 Better Use of Infrastructure

When a production system is not operating at an acceptable level, or is no longer competitive in the market, one option is to abandon the process, and build an entirely new production capacity. Nowadays, this is typically last resort and more often than not such an investment is postponed till the present infrastructure has reached its end-of-useful life. Until then, maintenance, and operational improvements of the existing infrastructure are cheaper options (actually most likely end-of-useful life may be defined as the point in time where the latter options become more expensive than building the new infrastructure from ground zero).

Investing in maintenance and improved operations is very much a control approach to system use. Normally this is the economic solution, as the cost of retrofitting or upgrading a sophisticated control system is typically much less than that of replacing the production system (not in the least because the latter will also require a sophisticated control system). Moreover a sophisticated control system yields information about the process behavior that identifies where investment in new infrastructure, or maintenance is the most beneficial from an overall economic point of view.

---

[3] The first and second law of thermodynamics inform us that no process can avoid degrading the environment. The total amount of energy in the universe is fixed (1st law), and any use necessarily degrades its quality (2nd law).

The effectiveness of investing in (more sophisticated) control will depend largely on how much room there is for improvement in exploiting the present plant behavior, both from a technical as well as economic point of view. So, it should be considered:

- How efficient (from an energy and/or a material flow point of view) is the present system? Is there enough room (in an economic or technical sense) for improvement?
- How flexible is the infrastructure to allow for system behavioral changes? How expensive is it to adjust the infrastructure, or retrofit sensors and actuators, to enable an improved economic return?
- How far is normal/desired system operation from the safety envelope? Is the best plant behavior under control, and/or sufficiently repeatable to exploit the difference between present operational conditions and the safety limit?
- How well is the plant known? Have we learned more about the plant since it was designed to allow perhaps for new ways of operating the plant, even beyond it present safety envelope?

11.3.5 Enabling Behavior

In some circumstances, plant and control design are integrated to achieve behavior that is entirely unnatural (unstable, but not impossible) for the plant to exhibit without control.

For example a forward swept wing ultrasonic aircraft is highly maneuverable but unfortunately its flight is unstable, and cannot be piloted without computer assistance. The slightest imbalance in lift between the wings destabilizes the flight path. It requires constant feedback to accomplish a normal forward level flying condition. No human pilot can fly such an aircraft without the intervention of an auto-pilot that constantly adjusts the wings so as to maintain the desired flight path. The advantage in realizing such an aircraft is of course unprecedented maneuverability and responsiveness. That there is no such plane at present indicates that the design task is hard.

Similarly, and less dramatic, if a bicycle had the back wheel as the wheel influenced by the steering (not a trivial mechanical exercise to complete) it would be rather difficult to ride the bike without falling. An autopilot, interpreting the desired human steering command, and correcting the steering angle as required to maintain the desired course can make all the difference.

A well-known example is the inverted pendulum. The natural position for a pendulum is to hang as gravity demands, nevertheless, with some control and feedback it is possible to maintain the pendulum balanced in the unnatural upward equilibrium. The control problem we all learn to master: walking in the upright position, rather than crawling on all fours. It is precisely this control task that makes it possible to launch rockets into space. Rockets are indeed mechanically similar to an inverted pendulum.

In biotechnology processes, high yield and high product quality are often mutually exclusive apart from some naturally unstable equilibria. Control can drive a system to this unstable equilibrium and maintain the process in this condition so as to achieve the desired outcome of a high yield and high quality product.

Feedback's Economic Benefits

In the engineered world **control** and **feedback** deliver economic benefits that are justified in terms of their capacity to deliver novel functionality by operating the plant and/or processes close to their natural safety envelope or even in an open-loop unstable situation.

In the latter case, the overall system operation and its safety is entirely dependent on the control and plant subsystems interacting as designed.

More typically control delivers functionality in terms of allowing higher complexity, greater safety and reliability as well as consistency of quality and efficiency.

11.3.6 Some Application Areas

It comes as no surprise that control is present in all industrial and agricultural activities, and more and more applications in the services-based economy rely on control and feedback. The following list is by no means exhaustive, but provided to illustrate somewhat the ubiquity of control and feedback in economy.

- *Agricultural industry*: micro-climate control in greenhouses, fertilization, irrigation systems, land surface leveling, automated harvesters.
- *Automotive industry*: robotic assembly lines, quality control, standardization of automotive parts, software systems, computer assisted braking, computer controlled traction, computer controlled stability, automated response to light conditions and rain conditions, in-car climate control, in-car voice activated functionality, engine control, exhaust and emission control, efficiency and fuel consumption control, autopilot parallel parking functionality, and soon to come autopilot services driving passengers from A to B.
- *Chemical and pharmaceutical manufacturing*: regulating process parameters, product yield optimization, satisfying safety requirements, minimization of energy consumption, maximization of efficiency or throughput.
- *Pulp and paper*: control to minimize raw materials, water and energy input, control to minimize waste products, control to improve the uniformity of paper quality, preventive maintenance of plant equipment, improving longevity of plant equipment, automated recycling of rejected paper, use of recycled paper in the material stream, speed and force control of the paper mill, improving the throughput (increasing speed without reducing quality).
- *Electricity generation, transmission and production*: control is essential to improve emissions and fuel efficiency in generation, maintain electrical parameters (voltage and frequency) within the preset range, and to improve grid size (national grids) for safety and reliability (an acceptable electrical grid has an up-time of more than 99.9%). To improve the penetration of alternative and more sustainable electricity generators (wind and solar) new control strategies will need to be developed as distributed generation is very different from the present operational condition characterized by distributed load and point generation. Control is also used to improve transmission efficiency by appropriately matching electrical supply and electrical demand.

- *Banking and fund management*: risk management or risk control of a portfolio, exchange rate speculation, portfolio asset allocation and even credit card fraud are approached using control and feedback ideas.
- *Defence*: unmanned vehicles executing autonomous missions of surveillance, air-to-air and ground-to-air defensive response, smart missile control, self guided and smart bombs, ship motion independent platforms for ballistic missiles, auto-pilot landing of helicopters onto a ship, and sensor network surveillance.
- *Disaster recovery*: unmanned vehicles executing autonomous or supervised search and rescue missions in situations too dangerous or too risky to deploy human rescue teams.
- *Process re-engineering*: such human and largely variable processes as passenger throughput in an airport, queuing in banking and government service centers, or even the triage of patients in a hospital can all be improved through the judicious use of control and feedback ideas.
- *Smart infrastructure*: actively controlled suspensions of large buildings, sway minimization of towers using water tanks and pumping to distribute water to minimize movement, lift distribution based on usage patterns, tension rod control to mitigate earthquake effects and distribute energy in an active manner in a building, people activated lights and heating that is adjusted for heat load, seasons and usage patterns, automatically shading windows, and motorway access control through price based on traffic density.

11.4 Societal Risks

As control subsystems become more and more pervasive our engineered world depends more and more on these control subsystems functioning properly. Failure in the control subsystem often leads to failure of the entire system. That should not surprise us. In modern medicine end-of-life is equated with (total) brain failure. The control (sub)systems act indeed as the engineered analog of the nervous system.

All our infrastructure such as oil refineries, water supply networks, electricity generation and transport systems depend on effective and continually operational control systems. Moreover, increasingly they are even interdependent and even openly connected through the internet or an intranet like communication infrastructure for monitoring, accountability and even supervisory control. Supervisory control and data acquisition (SCADA) systems are indeed common in all industrial control systems and they literarily control the operation of our environment on which we depend to conduct our daily lives.

As far as these control systems are protected from misuse and external malicious intrusion everything is fine, but unfortunately in general, no human engineered system is totally secure. There are no total guarantees.

Moreover security and accountability are in this case increasingly difficult to balance. Special care has to be taken to achieve both, and again there are no fail proof systems. We need to accept some risk. The trade-off between the benefits of pervasive control (the internet of things that can deliver the most sustainable society), the open availability of (selected and processed) information at different locations by a collection of different users (to provide the necessary checks and balances) and the inherent risks associated with the abuse of this information must be reached and legislated, as well as policed to ensure continuous safe operation of our engineered world.

11.5 Comments and Further Reading

The role of adaptive control in hemodialysis is described in Sternby (1996). A relatively up-to-date account of where bio-inspired feedback applied in hemodialysis is at present is Santoro et al. (2008).

The enormous success of the cochlear implant, due to advances in signal processing, feedback and microelectronics, but above all, due to the genius and tenacity of Graeme Clark[4] is described in Clark (2000).

Applications of robotics and haptic interfaces to train surgeons are pursued at a number of places, see for example the MUVES project at

http://www.muves.unimelb.edu.au/

Active noise control or active noise cancellation is described in relatively simple terms in Snyder (2000). It is a very pervasive technology.

The advent of domotics, home automation and ambient intelligence are apparent, and a number of books discuss the trends or introduce you to the wonderful or not so wonderful world of the robotic and the automated home (Soper 2005; Briere and Hurley 2007; Karwowski 2006).

It may come as a surprise, but backwheel steered bicycles have been around for a while, at least since 1869, but as we know, there are not many on the road. A website devoted to this amusing challenge is *http://www.wannee.nl/hpv/abt/e-index.htm*.

A brief history of control and feedback is found in at the beginning of Lewis (1992) and Franklin et al. (2006), which are both popular texts introducing feedback control to engineering students.

It appears that a history tracing the impact of feedback from the dawn of the industrial revolution until today is waiting to be written.

An overview of how pervasive SCADA is, and what the modern vulnerabilities are is presented in Shaw (2006). SCADA is also discussed in Lipták (1995). New standards for wireless networking and control networking are still being developed, see for example *www.isa.org* for more on standards in control and automation.

---

[4] Clark, Graeme, 1935–, Australian pioneer of the multi-channel cochlear implant. Founding director of the Bionic Ear Institute.

Chapter 12   A Look at the Future

12.1 Introduction
12.2 From Analog Controllers to Distributed and Networked Control
12.3 From Automatic Manipulators to Humanoid Robots
12.4 Artificial Intelligence in Control
12.5 Systems and Biology
12.6 Comments and Further Reading

Chapter 12
A Look at the Future

> *With high hope for the future,*
> *no prediction is ventured.*
>
> Abraham Lincoln

12.1 Introduction

Just from what we have seen so far, it transpires that the opportunities for feedback and control to touch all aspects of our daily lives are truly immense. Moreover, it is also true that we certainly have not yet exhausted neither theory nor innovation in control and feedback. In this last chapter, some of the recent advances and promising trends are sketched, allowing us to venture a glimpse into what may be. Certainly, we will have overlooked many advances. Paraphrasing Abraham Lincoln, we have great hope for the future of control, but realize all too well that our sight of the future is limited by ourselves. Our final comments and examples are offered to simply keep our minds open, and encourage us to participate and continue to share the joy of feedback.

This final chapter begins with advances in networked and distributed control, touching on ideas from cloud-computing and ambient intelligence. Next we look at automation, and in particular what robotics has to offer, an aspect of automation that we may have underplayed in the book so far. The connection with artificial intelligence and agent-based systems, topics of great research value in the computer science community is clear. Finally we look at biomimetics, and the promise of designing and manufacturing with biological systems just as we do now with electronics and mechanical systems.

12.2 From Analog Controllers to Distributed and Networked Control

The main idea of a control system is captured by Watt's governor: a (mechanical) device that is attached to what we desire to control (the speed), operates continuously (closing or opening the steam admission) so as to meet our objectives. In fact Watt's governor is often used as the symbol for the industrial revolution. Indeed, the steam engine would not have worked without it, and without the steam engine there would have been no industrial revolution. In some contexts, the governor is still seen as "advanced engineering". Certainly, as control engineers we are fond of the governor, but it is clearly a symbol of yesterday.

Today the controller is no longer a specific device devoted to perform a specific task. Digital control is implemented through a simple piece of software, operating on a generic computing unit. Any one control task shares the same physical resources (compute capacity, communication capacity, memory capacity) with many other tasks, communication, other control, coordination, reporting, interface requirements and so on. Moreover, control systems are no longer collocated with the process they manage. Controlled systems are distributed over space, and like the internet or a powergrid these systems can expand over the entire globe, or even beyond as is the case in an unmanned space mission. Just like actuators and sensors impose physical limits to what control can achieve, so do the limited resources of computation and communication. Designing control over a distributed sensor/actuator network with distributed communication and computational resources brings new challenges and opportunities. One of the fundamental challenges in such systems is related to our notion of time. We measure time using clocks, clocks that now are distributed over space, and hence there are many local versions of time and more importantly there are many different notions of time all in the same system. How is this all coordinated? How do we describe dynamics over multiple time axes, which perhaps are conflicting at a global level (but locally in space consistent)? In this sense, we can find analog time for local dynamics, periodic clocks (perhaps badly synchronized), and event-based timing. Does it matter if there is no such thing as a universal time axis? Much of our design tools are being adapted to deal with this very hybrid world, where in the end synchronization only occurs through the exchange of marked events.

12.2.1 Embedded Control Systems

Already applications have emerged where control is fully distributed and totally integrated in the process to be controlled. A modern day high density, nano-technology computer chip or a modern car are nice examples of this trend.

Refer to Fig. 12.1. There is servo braking (even brake-by-wire), computer assisted steering, traction control systems, active suspension, climate control, audio control and sound equalization, car lights responding to ambient light, windscreen wipers responding to ambient conditions, invisible engine control, fuel efficiency and pollution control systems and the trusted cruise control. Some cars offer collision avoidance systems and automated parallel parking. In some cities, cars communicate within a travel network and their global positioning systems (GPS) are coordinated to avoid congestion on the roads and produce in real time shortest travel time routes to your destination.

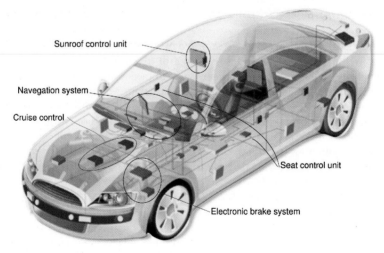

Fig. 12.1. A car with multiple distributed sensor

Really, why can your car not drive you home from work?

The control, computer and communication network infrastructure in a car is substantial. Already it may be viable to consider an internal wireless network rather than a wired network, purely from a fuel efficiency point of view (less cables = less weight = more fuel efficiency).

Even a cellular phone is a rather remarkable computer system, with enormous potential in a control context. For example tracking all phones', or tracking global positioning systems' information, and linking it into a traffic network information system could provide information about drive times, optimal routes, and in fact lead to a traffic control system that could regulate the traffic lights, provide individual route advice, change in real time speed limits (the potential exists to impose variable speed limits by communicating with engine control units, or vary the access fees to the road on the basis of whether or not the global advice is followed or not), and even schedule emergency vehicles. Using such technology one could envisage a user pay system for road access based on position information and driving behavior. There is an entire new discipline of *spatial information sciences* being developed around spatially enabling services.

In such large scale applications, the constraints are a large part of the design. New design questions are emerging on how should the available energy be used and distributed between sensors, actuators, communication, and computation? How much energy is required to manage an entire city's traffic network? How much more energy efficient can an automated, feedback-based, traffic network be versus the present policy-based or open-loop managed network?

Embedded control systems design requires the consideration of facts like

- the overall system will operate across many different modes, the various different subsystems will each have their own local goals to meet, and some of these local goals may even be competitive. Notions of global optimality vs local optimality and local fairness (as in fair access to information, and other resources at the local level in view of competition) will need to be addressed;
- the communication network must handle a variety of protocols and data. Moreover, network management is a sub-task in the overall system management. Communications must be allowed to be broken and re-instated. The network topology is ad-hoc;
- some part of the system will fail, functionality must be guaranteed, there must a graceful degradation of performance;
- available power and resource scheduling;
- share resources among many tasks, allow prioritization, and re-prioritization of tasks, fairness and access to resources. (Fit for purpose, or maximize revenue?)
- the potential to adapt and learn is large;
- maintain security and safety, alert when problems emerge, or may emerge;
- systems should be allowed to grow, slowly expand, even contract as to meet demand.

An aspect of increasing importance for all such large scale control systems is how does the system interface with the human society it is supposed to service? What is this human interface? How do we interact with the technology? Issues of privacy and personal security, the potential conflict between an individual's rights and expectations versus the rights and expectations of the society as a whole must be thoroughly considered.

12.2.2 Networked Control Systems

In networked control systems, each node in the network (later on we mention the concept of an agent) is able to perform some activities and it has access to local resources and most likely has local rules, and local goals. Coordination over the network enables the realization of network wide goals. The scenario is illustrated in Fig. 12.2, where only a small number of nodes has been identified, but the entire cloud can be thought of as consisting of similar nodes.

The picture is as follows. Each process has its own local sensor and actuation systems, actually its own version of a network, which is in itself networked with the amorphous cloud of networks. The term *federated* sensors and actuators is used to describe the situation. The idea of a network of things then emerges: a network of networks, and hence the potential for processes to share information, computational resources, and hence the ability to collaborate. This in itself poses new control questions: how is collaboration, monitored, encouraged, guided?

Questions like *who trusts who?* appear. What do we gain by collaboration? What do we loose? How is this internet of things working for society? A lot of research activity is devoted to this.

From a control and feedback point of view the issues are, complexity which is aligned with hierarchical model building and the degradation of the quality of information in the network, in particular due to delays, missing and potentially conflicting information as well as poorly timed information. This general information degradation is countered by the plurality of sensors and increased redundancy so that soft sensing can potentially overcome these problems. Data synchronization is a non-trivial problem in any such network.

This internet of things can be driven to the extreme. Imagine that your smart T-shirt (your computer interface), and your hearing aid, your shoes and watch, as well as the refrigerator at home and the milk bottles all participate in the internet. Imagination is free. Your vital signs are monitored, and your doctor provides advice that all is

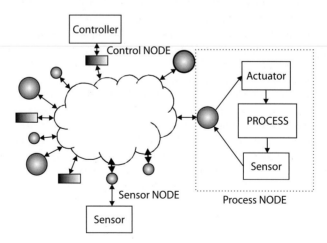

Fig. 12.2. Networked Control Systems

well, but that a little more exercising is necessary. A new software for your hearing aid will be installed to improve classical music perception, as you requested. Your attention is drawn to your upcoming schedule of meetings, and that a few participants will be late by about 10 minutes, as they are held up in traffic. You acknowledge. You open up a communication with mum, discussing dinner, and the refrigerator (which is listening in) intervenes informing you that you have run out of milk, which by the way is available from the shop just around the corner from where mum is now …

Such dreams lead us to the topic of cyber-physical systems.

12.2.3 Cyber-Physical Systems

A Cyber-Physical System can be defined as a system composed of sensors, actuators and computing components interacting with the physical world in real-time. It has all the characteristics of a distributed, resource-constrained, control system.

Direct information about the physical world is acquired through a number of local devices, which are spatially distributed and are often of a heterogeneous nature. They sense different things (temperature, light, speed, contact force, sound, chemicals), with different accuracy and range and different sampling techniques (and presumably different clocks). On the other hand, the operation and performance requirements are both local and global. Each element has its own objective but there are aggregate properties of the overall system. Thus, a cyber-physical system requires the integration of this heterogeneous physical layer and a global decision and control network, operating through decentralized and distributed local sensing/actuation structures. One of the great difficulties in a network of this kind is to interpret the sensor data, and even identifying their correct chronological order can be a difficult task.

Applications and instances of cyber-physical systems emerge in many diverse technological areas, including utility networks, cars, aerospace, transportation networks, telecommunication networks, environmental monitoring, biomedical and biological systems (like a body network), as well as in defence systems.

Autonomous unmanned craft control systems are now capable of managing a flock of autonomous unmanned aircraft/submarines/cars to execute a cooperative task lasting from a few hours to days. All the autonomous vehicles in the flock have local navigation, guidance, fault detection/recovery and data processing capabilities. They collaborate through the exchange of data, in particular sharing sensor information. Mission control optimizes trajectories to limit time and fuel, avoid inclement ambient conditions, and maximize the utility of the mission. A cyber-physical system should be able in the future to carry on the tasks assigned to it in a changeable scenario. They will interact with other active or passive elements and through feedback optimize the utility of their resources, for example enlarging the life span. We have seen that feedback is a fantastic way to deal with process uncertainty. But in this case, the uncertainty is extended to many other issues and involves decisions about immediate goals, use of resources, agents to deal with or information to process among many other.

In most (foreseen) scenarios we see autonomous elements (vehicles, robots, people) interacting in unstructured environments receiving information from both on-board sensors and networked sensors in the environment, exchanging data with other surrounding agents

Fig. 12.3. Cars running in convoy at high speed

and performing some predefined tasks. An ongoing project (PATH[1]) improves highway utility by means of network coordinated driving of vehicles, as illustrated in Fig. 12.3. The ideas behind *Better Place*, a company that targets the reduction of oil consumption in transport, rely very much on feedback and control over networks to achieve their goal.

It is difficult to understand how to translate complex human intent into priorities and physical actions even for a single cyber physical system interacting within a known scenario, let alone with an unstructured one, full of competing priorities. Again in these developments CCC (computing, communication and control) will need to come together in a more fundamental way than hitherto has been the case. Indeed in order to realize Norbert Wiener's original dream of cybernetics of true autonomy in the machine environment, a machine environment that interfaces seamlessly within a human society, and indeed serves the latter, control needs much further development.

Among the main challenges in designing such cyber-physical systems we see:

- efficient, partial spatial-temporal-physical modeling of a system and estimating the validity or uncertainty in and of such a model;
- selecting and optimizing relevant objectives, amongst a set of potentially incompatible objectives;
- ensuring safety at a local and global level;
- coordination of local activities, to participate in global activities, within resource limits.

12.3 From Automatic Manipulators to Humanoid Robots

The industrial revolution, at the turn of the 18th century, freed humans and animals from the most physically demanding activities, using the steam generator as the engine of choice (see Fig. 8.1). The advent of computers in the middle of the 20th century determined a tremendous advance in data processing, and started to relieve humans from some mental activities, where the machine is faster, more reliable, and more enduring. The relentless drive to minimize physical activities continues unabated, and most repetitive actions are now supported or programmed into automated systems.

---

[1] Partners for Advanced Transit in Highways, has got a number of developments in this direction, see http://www.path.berkeley.edu/PATH/General/.

The combination of powered systems with data processing and automation in manufacturing has been and still is one of the main drivers for feedback and control in engineering. Robotic manipulators are a natural evolution in this line. As always the aim is to manufacture with greater speed, using less natural resources, less energy and power and yet improve product quality. Assembly-disassembly activities are mostly automated and some industries, like car manufacturers, extensively use robots to cover all routine activities.

Robots provide greater flexibility in manufacturing cells, without sacrificing manufacturing speed and accuracy. They are particularly useful for highly repetitive tasks, where human attention span quickly becomes a major issue. From a control viewpoint, robots have provided new challenges to study. They are intrinsically non linear, as robot dynamics can be described in a very elegant manner using an Euler-Lagrange formalism, mainly due to links interactions, fast and complex, forcing us to deal with highly nonlinear system analysis and synthesis. In the last years it has deserved a huge amount of research (Ortega et al. 1998). Collaboration of many robots in a single manufacturing cell is quite common.

Nevertheless, despite their usefulness, and reliability, present day robots are as yet not at a point of engineering evolution that you want one of them for doing the garden work. Moreover, in comparison to humans, they are also quite expensive from an energy point of view.

12.3.1 The Humanoid Challenge

The quest to create a humanoid robot is well under way.

For example, the robot soccer challenge (RoboCup) is developing a team of autonomous robotic players able to beat the human world champions in a game of soccer, using the official FIFA rules before 2100. This is a lofty goal which is perhaps even misplaced. As clearly once a robotic humanoid is developed that will play a game of soccer on its own energy source, and has *no* mechanical or other capabilities that outperform a human soccer player, and one that never trespasses the rules of the game, well then there is no reason to believe that they indeed can win, or should win, unless there is a winning strategy for soccer which is rather doubtful. The emphasis on no capabilities greater than a human is ours to bring into focus the question about what does humanoid actually mean? Once we can build human-looking robots, why would we stop at human capabilities? A titanium frame will do nicely, thank you. Clearly some serious ethical questions have to be addressed before we enter such an era of technological development. A little closer to home, a simple robot to mow the lawns, that does not scare grandma, is friendly with the dog, and knows the difference between a tulip and grass, and informs us when it needs maintenance (but no more than once a year) would be great.

The real challenge here is the Turing test.[2] It is very conceivable that machines and/or robots will be able to pass for all practical purposes the Turing test this century. At this time we will have the ability to interface with robots in a natural manner.

Will it really help to make Earth a better place? Perhaps we could start trying to live in peace with ourselves and our neighbors before we learn to deal with robots as well?

---

[2] Alan Turing describes this in his 1950 paper "Computing Machinery and Intelligence". A machine passes the Turing test if it is not possible for a human using everyday language as the interaction medium to tell reliably the difference between a human or machine interactor.

12.3.2 Master-Slave Systems

Robotic systems that assist humans in search and rescue operations, or complete difficult surgery in remote places (like in a space mission) are clearly more within reach. In fact there is no obvious impediment to develop robotic surgery to a degree of sophistication that is more reliable and more accurate than what could be performed by human hands: in particular when micro-surgical equipment is involved. Up to now, most of such activities are open-loop supervised with a human as the last element in the loop. In this situation, it is the human whose actions are copied by the remote machine as in a master-slave system, although due to intrinsic delays in tele-operation and in order to achieve smooth motions a number of local actions are fully automated. It will be a while before we see robots perform surgery at the mere command of a medical doctor, or see an autonomous robotic search and rescue squad rescue people from an earthquake devastated zone.

More in our immediate future, robotic hands, feet and legs will improve the quality of life for many amputees. The main open issues relate to how to provide a natural interface, so that the brain or nervous system can be seamlessly in control of the artificial limb.

12.4 Artificial Intelligence in Control

The development of autonomy in robotics as well as cyber-physical spaces leads naturally to the research topic of artificial intelligence, a topic founded in earnest by Alan Turing. This is an interesting development as the understanding of natural intelligence was and still is one of the motivating quests of cybernetics.

In control this trend is seen in the pursuit of so-called *intelligent control*. The terminology is unfortunate and somewhat controversial as it creates the impression that other control approaches are perhaps not intelligent.

Ideas from Artificial Intelligence (AI), that is techniques based on machine learning and machine reasoning, have proven useful both in theory and practical applications. Machine learning has a significant body of work dealing with the notion of *complexity*, which is very useful in understanding at least in principle the difference between what is a tractable or a non-tractable problem. Machine learning approaches to both modeling and control are under development. Successful applications have been reported in particular in the control of processes that are poorly quantified, and where experience and qualitative descriptors play a dominant role.

Neural networks, fuzzy logic, expert systems, evolutionary computation are some of the ideas developed in this general area of research. There are interesting results and some nice niche achievements, but the theory and even the practice is still in its infancy. It is fair to say that we have a long way to go before we have a computational environment that exhibits flexible and reliable learning (that can summarize and structure experience) and autonomous decision making that can deal with unstructured and novel environments.

Expert systems is one of the first developments in artificial intelligence. Using logic reasoning and decision trees expert systems are able to solve a number of interesting problems. They provide excellent tools in capturing human expertise in a systematic manner and can be used to assist human reasoning. For example the triage problem of patients in a large scale emergency situation can be greatly assisted using an expert system to guide less experienced health professionals. They are also often used in assisting in the training of new staff, for example in training maintenance staff for complex equipment, or in training staff through emergency processes where they are often used in conjunction with simulated scenarios. The tools to capture human experience, and build complex expert systems are still rather limited, and many of the original research questions, in particular around automated reasoning and the discovery of new knowledge remain largely unresolved.

Artificial *neural networks* have found significant application in modeling highly nonlinear processes, in a compact manner. They are equally useful in fault diagnostics, in particular when fault characteristics and alarm conditions depend on a highly nonlinear manner on a multiplicity of observed quantities. There is significant literature dealing with the theory and application of artificial neural networks. The IEEE Neural Network community defines its field of study as *the theory, design, application and development of biologically and linguistically motivated computational paradigms involving neural networks, including connectionist systems, genetic algorithms, evolutionary programming, fuzzy systems, and hybrid intelligent systems in which these paradigms are contained*[3].

Fuzzy logic or *fuzzy systems*, considered a branch of neural networks, were developed to deal with approximate knowledge, qualitative concepts and even contradictory facts in a systematic way. Fuzzy systems make extensive use of set-theoretic notions, and build on concepts of probability or possibility theory (which is particularly tuned towards incomplete information), to represent uncertainty. Its intuitive approach has attracted a lot of attention. Fuzzy logic ideas are used in such consumer items as washing machines, dish washers and video cameras to provide enhanced functionality.

Techniques to support learning, modeling and control in neural networks use ideas from *genetic algorithms* and so-called *evolutionary programming*. At their core these are essentially optimization approaches, specifically conceived to deal with a large number of variables, exploiting a specific class of nonlinear models which are known to have good universal approximation properties. At an intuitive level they borrow ideas from biological *evolution* theory.

A good overview of the basic theory of machine learning placed within a context of control and systems engineering can be found in Vidyasagar (1996).

Despite the plethora of applications and the progress made to date, artificial neural network theory and practice is still far removed from the goal of delivering a truly autonomous, artificial brain-like device capable of learning both in supervised and unsupervised environments, able to reason and discover new knowledge. In our opinion, this end-goal is not achievable without much more progress in unraveling the functional structure of the human brain. From a device point of view, (super)computers can already be constructed with a component count (a few 100 billion transistors is not inconceivable) that rivals the cell count in a human brain. Moreover these electronic devices can switch at a million times the speed our neural cells can fire, and in prin-

---

[3] Quoted verbatim from the IEEE Transactions on Neural Networks www-site.

ciple they can do this in a sustained fashion over a period spanning many years. Where the electronic world is still well-behind is in raw input/output capacity (both multiplicity of sensors, as well as raw input bandwidth, which in the human brain can peak at 10^{18} switch events per second, an estimate derived from the number of input synapses into the cortex) and in the specifics of actually learning and storing knowledge. Power consumption and packing density are other aspects where the physical differences between the artificial and human brain are immense, with the latter clearly a few orders of magnitude ahead. There is a way to go, but it is not inconceivable with the amount of research effort being devoted to understanding our brain. It is after all the frontier of mental health that is challenging us most, and with a steady progress in cloud computing and stream computing, we will see artificial brains with a capacity and functionality like the human brain by the end of the 21st century.

12.4.1 Ambient Intelligence

The European Union IST Advisory Group report of 2003[4], defines "an Ambient Intelligence Environment" as an infrastructure environment of sensors and actuators overlayed with intelligent interfaces supported by computing and networking technology that is embedded in everyday objects such as furniture, clothes, vehicles, roads and even building materials, like insulation material, but even particles of decorative substances like paint. Ambient intelligence envisages an information environment of seamless computing, advanced networking technology and human interfaces. Such an environment is aware of the specific characteristics of human presence and adapts itself to the expectations of the users of that environment. It is capable of responding intelligently to spoken or gestured indications of desire. One even envisages systems that are capable of engaging in an intelligent dialogue with the users when the need arises to reduce uncertainty or ambiguity. In general though, there is an expectation that the users are entirely oblivious to the ambient intelligence.

In ambient intelligent environment research and development it is natural to focus on the enabling technologies: capturing, transmitting and processing information. The real advances are however in the way the system is able to interpret data and react. This is control and feedback. Uncertainty, changing scenarios, conflicting and time-varying goals as well as resource limitations will be some of the issues that must be dealt with. The whole concept of autonomy and seamless interaction with humans begs for questions of safety and hierarchy as well as who actually is in control?

Ubiquitous computing, communication and control are the key ingredients with which to implement ambient intelligence. An ambient intelligent environment is considered to be a step up in complexity from networked control systems, because of the distributed nature of environment-human interaction. Nevertheless all the problems that exist in the simpler networked control environment, with respect to spatial and temporal time scales, degradation of information and coordination of resources exist here too. Moreover, the computational resources required to interact with human voices,

---

[4] ISTAG coordinates its information and communication technology research on behalf of the European Union *ftp://ftp.cordis.europa.eu/pub/ist/docs/istag-ist2003-consolidated_report.pdf.*

gestures, and behavior more generally, presumably across a variety of languages and cultures are for the present unimaginable outside a human society and even there we are not doing too well either. If we cannot construct a socio-cultural environment that works conflict free for humans, how do we expect to create one for humans and machines?

The idea of *agents* or (software) individuals, is being used as the way forward in ambient intelligent environments.

12.4.2 Agents

An *agent* is an autonomous entity with limited knowledge of its environment, has a capacity to react to this environment-based on a combination of predetined rules and past information received from the environment, which includes possible interactions with other agents in pursuit of a goal on behalf of a user, another agent, or itself.

This minimalist definition is rather general. As a consequence the class of agents is perhaps hopelessly broad and includes all animal and plant life, but also such simple devices as sensors and cars. Also an industrial organization and even the human society considered as a whole can all be viewed as special instances of agents.

What is expected from an agent by acting in an intelligent way is to do what is appropriate for its circumstances and its (present) goal, being flexible (adaptive) to changing environments and changing directives, learning from experience, and making appropriate selections based on the available information.

Multi-agent systems is an approach to adaptive control across a distributed system rooted in computer science. It is one of the fastest growing and promising research areas in computer science. Multi-Agent Systems constitute a very interesting approach to deal with complex systems, either in a simulated or virtual environment but also to control and maintain industrial environments.

A very interesting line of research has to do with the characterization of the behavior that emerges from a collection of agents working together. The group behavior even from a collection of very simple agents can be incredibly complex and powerful, as for example borne out in such applications as flocking and follow-the-leader. A follow-the-leader implementation, where each agent has to follow a leader (agent) is presented in Fig. 12.4 consider the similarities with Fig. 12.3), here all the agents have the same goal and they need to reach a consensus among them in that they have to achieve the

Fig. 12.4. A flock of birds (or hordes of devices?)

same heading angles and maintain a safe separation distance. Simple feedback mechanisms are at work, but our understanding of such dynamics is still rather primitive, and there is a long way to go. One particular line of investigation is based on observing herd behavior and flocking in the animal world. The simplicity of the rules that lead to the fascinating behavior of schools of animals is truly remarkable and inspiring.

12.5 Systems and Biology

Life and nature have always been an inspiration for the creation of engineered systems.

Flight is inspired by birds. Certainly aircraft are by no means as elegant as birds, but we learned to fly a lot faster and higher.

Velcro, invented by De Mestral, is based on a natural phenomenon of burs sticking in clothes and hair.

The recent explosion in the quantitative study of biology, and the variety of detailed mathematical models that become available describing nature from the sub-cellular processes surrounding protein folding to entire organs present new opportunities and challenges for systems engineering.

12.5.1 Bio-Systems Modeling

The issue in modeling biology is really one of scales.

Protein folding at the basis of cellular processes requires a quantum physical description of the behavior of atoms and forces, working on time scales of fempto seconds 10^{-15}, on a spatial scale measured in nano meter 10^{-9}, whereas a typical human being may have a life expectancy of around a billion seconds 10^9 and is measured in meters. Modeling over 26 orders of magnitude in time, and 9 orders of magnitude in space is extremely challenging.

From a feedback point of view, the question is really, when is it important to have information at the nano scale level to effect the unit scale behavior? It is not conceivable to maintain all nano scale information at the same level of accuracy on an ongoing basis. Also our own experience indicates that it is not necessary in general, but when does it matter to know?

Feedback may come to the rescue here. Think of the operational amplifier. Just as high gain allows the decoupling of systems, so can scales be aggregated and summarized in constitutive laws, that only need to be unraveled when necessary, and only then. Successful examples of this philosophy are emerging, for example the comprehensive model of a human heart the so-called Auckland model at the University of Auckland's Bioengineering Institute lead by Professor Peter Hunter.

As successful models are at the basis of successful feedback design, there is a way to go before we can commence design ab initio in this arena. One of the grand challenges here is to develop a sufficient library of models, as well as a standard language to express these in, so that we can start building new models, but also start thinking about synthesis with biological building blocks in the mix amongst engineered units. The road ahead is in view, but we have just started. The "in silico" cell or bacterium is still a few years away. Development along these lines will require a truly world wide collaborative effort. It is well-worth the effort, as the potential is indeed enormous.

12.5.2 Biomimetics

Lacking full models and understanding does not stop us from mimicking partial behavior and leveraging this in an engineered environment (think of submarines, airplanes and velcro to name but a few objects).

To date the most successful biomimetic invention is most likely the digital audio encoding standard called MP3. In the MP3 standard sound is much more compactly represented than is traditionally the case in say the compact disk audio standard. The latter uses the Nyquist-Shannon criterion to represent sound in a nearly exact manner. MP3 however is a form of lossy compression, that is information is discarded during the encoding. It is based on the simple observation that there is no need to record or encode what our ears cannot hear. The MP3 sound is not the same as the sound that was generated in the first place, but our ears cannot tell the difference. It is easy enough to physically measure the difference, a microphone can hear the difference, but our ears cannot. This is so-called psychoacoustic or perceptual coding. For music it is perfectly acceptable to use MP3 encoding, but in other applications like acoustic diagnostics of helicopter gear boxes or acoustic diagnostics of the heart function lossy compression will simply not do.

12.5.3 Bionics

Bionics, the interaction of mechanics, physics and electronics with biology, relies heavily on feedback.

The bionic ear (Clark 2000), a multi-channel (around 20 electrodes at present is state of the art) electric stimulation of the auditory nerve, would still be a dream if it had to equal the stimulation (about 20 000 channels) of the healthy ear. The brain's learning ability, coupled with training (all forms of feedback) allow deaf people to lead practically normal lives using a device that has but the fraction of the stimulation capacity of the normal ear. Using greater understanding of how the healthy auditory nerves are stimulated (better models of the biology) further advances are expected in this technology, with the promise of substantially shortening the training phase.

There is not an organ in our body that is not targeted with some form of a *spare part* technology. Most of these benefit from feedback and control ideas, as after all the body parts they replace are performing their function because of feedback and control (interacting with the rest of the body). Pace makers must react to physical activity, and the artificial pancreas helps to regulate glucose in the body. Artificial hearts, lungs, stomach, are all available in one form or another, and it is probably safe to predict that a whole range of spare body parts will become available to assist us to live better and longer.

At present the interface between the artificial organ and the human body is still rather crude. Great advances are anticipated if a biologically compatible interface with nerves (control and command) and blood (for energy) could be created. Tapping into the nervous system or its signals for command, feedback and control, and extracting energy from glucose would make long lasting truly bio-engineered components. Again to realize this dream we need better models of the biology, and a better understanding of creating interfaces between biology and mechanical/electrical devices. The breakthrough we need is to understand the code which is used to represent information on the nervous system.

12.5.4 Bio-Component Systems

Going a step further, it is possible to foresee a future where biological cells and subcellular mechanisms will become integral parts of engineered systems. This requires much further developments of quantitatively accurate models of cells and cellular mechanisms, but conceptually from a systems engineering point of view there is no barrier. A box in a block diagram has a behavior, and as long as that behavior is characterized, it can be used for design. As soon as it can be interfaced (reliably) it can be synthesized, once it can be manufactured reliably (also for this we may adopt ideas from biology). So, given advances in systems biology (still a way to go), and advances in biochemistry to produce reliable manufacturing processes that interface with silicon (not trivial), and advances in software engineering (to deal with the complexity) it is very conceivable that a new generation of engineers will work with systems where silicon chips and cells and even organs are just some of the building blocks they work with to design systems. The need for such hybrid technology is of course to exploit the strengths of both fields: the extreme speed and stability of silicon, reliable and repeatable; combined with the energy frugal, and incredible sensitivity of biological processes.

Imagine creating a brain-like computer that delivers 10^{15} operations per second for less than 10 W of power (5 orders of magnitude better than present day computers) with the longevity, repeatability and accuracy of today's silicon-based electronic computers. With such devices, the Turing test will become a triviality.

12.5.5 Protein, and Nano-Scale Biochemical Engineering

Biochemical engineers are able to use cells and biology more generally as factories. The vision is to create new molecules, new drug delivery mechanisms, or even repair damaged tissue in situ in the body. Perhaps the technology could be developed to grow a new organ, like repair the cochlear, rather than relying on a bionic one. It is difficult to imagine what we can do when we understand biology to the extent that we can fully program DNA, and are in principle able to create new life forms, or regenerate old ones.

These protein processes work at a molecular level within sub-cell factories, that exploit molecular sensing, messaging and molecular feedback mechanisms that are being unraveled. These are sure to open up new questions but more importantly bring new understanding that we could use to implement feedback more effectively (using less energy for example) in complex situations. The research and development work in this area is truly fascinating, and promises access to a complexity of manufacturing we can only dream of at present. Drug delivery mechanisms as shown in Fig. 12.5, where the complexity of a protein and the shape of an "intelligent" tablet are shown. In such developments, feedback and control are purely mitigated through biochemical processes.

Clearly the potential of such a technology is enormous, but it also poses some serious ethical questions. Why stop at growing organs? Is it ethical to create an emotion-free ε-class biologically grown robot, just because we can?

Protein model: ~10 nm   Nano-capsule: ~100 nm

Fig. 12.5. Comparing a protein with an engineered nano-tablet

12.6 Comments and Further Reading

Information about Embedded Control Systems, and the many research activities related to embedded control systems, can be found on the webpage of the project funded by the European Union: *Network of Excellence ARTIST2, www.artist-embedded.org/artist/*. An interesting and modern introduction to the topic of embedded systems is Vahid and Givargis (2001).

Networked control systems is a very active research field at present. A www-site devoted to providing an overview of the activities in this field is *http://filer.case.edu/org/ncs/basics.htm*. A compilation that brings networked control and embedded control systems together is Hristu-Varsakelis and Levine (2005).

Ambient intelligence is discussed in some detail in Weber et al. (2005) and Remagnino et al. (2005). The latter focuses in particular on the issue of the human-computer interface. In general though, serious questions dealing with privacy, security and indeed ethics must be raised (Ahonen and Wright 2008).

The quest for the humanoid robot team is described on RoboCup^{TM}'s website *www.robocup.org*. A theoretical overview of human-like biomechanical modeling and control rooted in an Euler-Langrange formalism can be found in Ivancevic and Ivancevic (2006). A comprehensive simulation engine based on the ideas in this book has been developed.

The Turing test is explained in Turing (1992) and Moor (2003). The latter also reviews a less ambitious but related idea and the associated Loebner prize. This Loebner prize, a yearly competition in artificial intelligence, is described at *www.loebner.net/Prizef/loebner-prize.html*. Turing laid the foundations of artificial intelligence. The conflux of systems theory and artificial intelligence was really the motivation for Norbert Wiener's cybernetics (Wiener 1948, 1954). The ethical issues raised in the latter are more relevant today than ever. The link between systems theory, cybernetics and artificial intelligence and more general computer science with the human brain was laid by Von Neumann (1958).

The computer science of agents and multi-agents is presented in the classic Ferber (1999) and more recently expanded in Shoham and Leyton-Brown (2009). Advances in cloud computing and stream computing will play a significant role in this field.

Systems biology that feels like systems engineering, explaining ideas of feedforward and feedback (though not quite in the systems language) is described in Alon (2007). The number of texts in this rather young area is really astounding. The computational and scaling issues we alluded are discussed in Kriete and Eils (2006). For a demonstration of what can be done in terms of multi-scale modeling, the impressive work by Peter Hunter's group in Auckland on the human heart is definitely worth a visit *www.bioeng.auckland.ac.nz*. Creating interfaces with neurons, and interacting with the brain more broadly is the field of neural engineering. Brain wave human-computer interfaces, brain activated and controlled prostheses and a better understanding of brain behavior are part of the rich field of neural engineering. A text that brings control and feedback together with neuroscience is Eliasmith and Anderson (2004).

Bionics and more in particular biomimetics is the topic of Bar-Cohen (2006). The bionic ear story is told in Clark (2000).

Within this context of the future of systems engineering, control and feedback, it is worth reading Holton (1998), and remember that no true advance can be made without accepting the responsibility of one's actions.

References

Ahonen P, Wright D (2008) Safeguards in a world of ambient intelligence. Springer-Verlag, Berlin Heidelberg
Albertos P, Sala A (2004) Multivariable control systems: An engineering approach. Springer-Verlag, Berlin Heidelberg
Albertos P, Strietzel R, Mort N (1997) Control engineering solutions: A practical approach. The Institution of Electrical Engineers
Alon U (2007) An introduction to systems biology: Design principles of biological circuits. Chapman and Hall/CRC, Boca Raton, FL
Anderson BDO, Vongpanitlerd S (2006) Network analysis and synthesis – A modern systems approach. Dover Publications, New York
Anderson BDO, Bitmead RR, Johnson CR, Kokotovic PV, Kosut RL, Mareels IMY, Praly L, Riedle BD (1986a) Stability of adaptive systems: Averaging and passivity analysis. MIT Press, Boston, MA
Anderson BDO, Bitmead RR, Johnson CR, Kokotovic PV, Kosut RL, Mareels IMY, Praly L, Riedle BD (1986b) Stability of adaptive systems: Passivity and averaging analysis. MIT Press, Cambridge MA
Antoulas AC (2005) Approximation of large-scale dynamical systems. SIAM, Advances in Design and Control
Ash RB (1965) Information theory. Interscience, New York
Astrom KJ, Hagglund T (2005) Advanced PID control. ISA – The instrumentation, systems, and automation society, Research Triangle Park, NC 27709
Astrom KJ, Murray R (2008) Feedback systems: An introduction for scientists and engineers. Princeton University Press, Princeton Oxford
Astrom KJ, Wittenmark B (1988) Adaptive control. Addison-Wesley
Bannister BR, Whitehead DG (1991) Instrumentation: Transducers and interfacing, 2nd ed. Chapman and Hall
Bar-Cohen Y (2006) Biomimetics: Biologically inspired technologies. CRC Press, Boca Raton
Barbera E, Albertos P (1994) Fuzzy logic modeling of social behavior. Cybernetics and Systems: an International Journal 25-2:343-358
Barbera E, Albertos P (1996) Psychological human behaviour: A system approach. In: Moreno Díaz R, Mira-Mira J (eds) Brain Processes, Theories and Models. MIT Press, pp 194-203
Belevitch V (1968) Classical network theory. Holden-Day
Bennet S (1979) A history of control Engineering, 1800-1930. The Institution of Engineering and Technology
Bennet S (1993) A history of control engineering, 1930-1955. The Institution of Engineering and Technology
Berka K (1983) Measurement: Its concepts, theories, and problems. Springer-Verlag, Berlin Heidelberg
Bode HW (1945) Network analysis and feedback amplifier design. D. Van Nostrand Co., New York
Bondía J, Moya P, Picó J, Albertos P (1997) MARCK: An intelligent adaptive real-time system for ceramic kiln control. La lettre de l'IA 123:260-267
Boring EG (1930) A new ambiguous figure. American Journal of Psychology 42:444
Boyd SP, Barratt CH (1991) Linear controller design: Limits of performance. Prentice Hall, Englewood Cliffs, NJ

Breedveld PC, Dauphin-Tanguy G (1992) Bond graphs for engineers. Elsevier Science Publishers, Amsterdam
Briere D, Hurley P (2007) Smart homes for dummies, 3rd ed. John Wiley & Sons
Brockett RW (1970) Finite dimensional linear systems. John Wiley & Sons, New York
Camacho EF, Bordons C (1995) Model predictive control in the process industry. Springer-Verlag, Berlin Heidelberg
Cannon RH (2003) Dynamics of physical systems. Courier Dover Publications, New York
Cantoni M, Weyer E, Li Y, Ooi SK, Mareels I, Ryan M (2007) Control of large-scale irrigation networks. In: Bailliuel J, Antsaklis P (eds) Proceedings of the IEEE, Special Issue on the Technology of Networked Control Systems, vol 95(1), pp 75–91
Cassandras C, Lafortune S (2008) Introduction to discrete event systems. Springer-Verlag, New York
Catlin DE (1989) Estimation, control and the discrete Kalman filter. Springer-Verlag, Berlin Heidelberg
Chen BM, Lee TH, Peng K, Venkataramanan V (2006) Hard disk drive servo systems, 2nd ed. Springer-Verlag, Berlin Heidelberg
Clark G (2000) Sounds from silence: Graeme clark and the bionic ear story. Allen & Unwin, St Leonard, NSW Australia
Colonius F, Kliemann W (2000) The dynamics of control. Springer-Verlag, New York
Cruz BJ, Kokotović PV (1972) Feedback systems. McGraw-Hill
CTA, Computer Technology Associates and Goddard Space Flight Center (1982) Integrated command, control, communication and computation system design study: Summary of tasks performed. Computer Technology Associates, Englewood, *http://nla.gov.au/nla.cat-vn4174102*
Daubechies I (1992) Ten lectures on wavelets. Society for Industrial and Applied Mathematics, Philadelphia, PA
David R, Alla H (2005) Discrete, continuous, and hybrid petri nets. Springer-Verlag, Berlin Heidelberg
de Melo W, van Strien S (1991) One-dimensional dynamics. Springer-Verlag, Berlin
Desoer CA, Vidyasagar M (1975) Feedback systems: Input-output properties. Academic Press
Doyle JC, Francis BA, Tannenbaum AR (1992) Feedback control theory. MacMillan, New York
Dropka E (1995) Toshiba medium PLC primer. Newnes
Eliasmith C, Anderson CH (2004) Neural engineering: Computation, representation, and dynamics in neurobiological systems. MIT Press, Cambridge, MA
Erickson KT (1999) Plantwide process control. Wiley-IEEE
Eurén K, Weyer E (2007) System identification of open water channels with undershot and overshot gates. Control Engineering Practice 15(7):813–824, *www.scopus.com*
Evans RJ, Mareels IMY, Sciacca L, Cooper DN, Middleton RH, Betz RE, Kennedy RA (2001) Adaptive servo control of large antenna structures. In: Goodwin GC (ed) Model identification and adaptive control. Springer-Verlag, London Berlin
Ferber J (1999) Multi-agent system: An introduction to distributed artificial intelligence. Harlow Addison Wesley Longman
Franklin G, Powell JD, Emami-Naeini A (2006) Feedback control of dynamic systems, 5th ed. Prentice Hall
Gawthrop PJ, Smith LS (1996) Metamodelling: Bond graphs and dynamic systems. Prentice Hall, Englewood Cliffs, NJ
Gevers M (1997) Communications, computation, control, and signal processing: A tribute to Thomas Kailath. In: Paulraj A, Roychowdhury V, Schaper C (eds) Modeling, identification and control. Kluwer Academic Publishers, Berlin, pp 375–389
Gislason D (2008) Zigbee wireless networking. Newness, Elsevier, Burlington, MA
Goldbeter A (1997) Biochemical oscillations and cellular rhythms: The molecular bases of periodic and chaotic behaviour. Cambridge University Press
Goodwin GC, Sin KS (1984) Adaptive filtering prediction and control. Prentice-Hall
Goodwin GC, Graebe SF, Salgado ME (2001) Control system design. Prentice Hall International, Upper Saddle River, NJ
Graham S, Kumar PR (2003) The convergence of control, communication, and computation. Proceedings of PWC 2003: Personal Wireless Communication, Lecture Notes in Computer Science, Volume 2775, Springer-Verlag, Heidelberg, pp 458–475

Green M, Limebeer D (1995) Linear robust control. Prentice Hall, Englewood Cliffs, NJ
Guckenheimer J, Holmes Ph (1986) Nonlinear oscillations, dynamical systems, and bifurcations of vector fields. Springer-Verlag, New York
Gyftopolous EP, Beretta GP (1991) Thermodynamics: Foundations and applications. McMillan, New York
Hailman JP (2008) Coding and redundancy: Man-made and animal-evolved signals. Harvard University Press, Cambridge, MA
Hannan EJ (1967) Time series analysis. Methuen, London
Haykin SS, Van Veen B (2002) Signals and systems. John Wiley & Sons
Hjalmarsson H, Gevers M, De Bruyne F (1996) For model-based control design, closed-loop identification gives better performance. Automatica 32(12):1659–1673
Hoagland M, Dodson B (1995) The way life works: The science lover's illustrated guide to how life grows, develops, reproduces and gets along. Times Books, New York
Holton GJ (1998) The advancement of science, and its burdens. Harvard University Press
Hoydas Y, Ring E (1982) Human body temperature: Its measurement and regulation. Plenum, New York London
Hristu-Varsakelis D, Levine WS (2005) Handbook of networked and embedded control systems. Springer-Verlag, Berlin Heidelberg
Hyland ME (1989) Control theory and psychology: A tool for integration and a heuristic for new theory. In: Hershberger WA (ed) Volitional action: Conation and control. Elsevier
Ifeachor EC, Jervis BW (2002) Digital signal processing, 2nd ed. Prentice Hall, Edinburgh Gate
Isidori A (1995) Nonlinear control systems. Springer-Verlag, Berlin New York
Ivancevic VG, Ivancevic TT (2006) Human-like biomechanics: A unified mathematical approach to human biomechanics and humanoid robotics. Springer-Verlag, Berlin Heidelberg
Janocha H (ed) (2004) Actuators: Basics and applications. Springer-Verlag, Berlin Heidelberg
Johnson MA, Crowe J, Moradi MH (2005) Pid control: New identification and design methods. Springer-Verlag, Berlin Heidelberg
Kailath T (1980) Linear systems. Prentice-Hall, Englewood Cliffs, NJ
Kalman RE (1960) A new approach to linear filtering and prediction problems. Trans. ASME (D): J Basic Engineering 82D:35–45
Karwowski W (2006) International encyclopedia of ergonomics and human factors, 2nd ed. CRC Press
Kawaguchi T, Ueyama T (1989) Steel industry II: Control system. Gordon and Breach Science Publishers, New York
Kay SM (1993) Fundamentals of statistical signal processing. Vol. I: Estimation theory. PTR Prentice Hall, Englewood Cliffs, NJ
Kay SM (1998) Fundamentals of statistical signal processing. Vol. II: Detection theory. PTR Prentice Hall, Englewood Cliffs, NJ
Khalil HK (2002) Nonlinear systems, 3rd ed. Prentice Hall, Upper Saddle River, NJ
Klaassen KS, Gee S (1996) Electronic measurement and instrumentation. Cambridge University Press
Körner TW (1988) Fourier analysis. Cambridge University Press
Kriete A, Eils R (2006) Computational systems biology. Elsevier, Amsterdam Boston
Kučera V (1979) Discrete linear control: The polynomial approach. John Wiley & Sons, Chichester
Kuhl J, Beckman J (1992) Volition and personality: Action and satate-oriented modes of control. Hogrefe and Huber, Göttingen
Kuznetsov YA (2004) Elements of applied bifurcation theory. Springer-Verlag, New York
Landau ID (1979) Adaptive control – The model reference approach. Marcel Dekker, New York
Lee EA, Varaiya P (2003) Structure and interpretation of signals and systems. Addison-Wesley, Boston
Lee WS, Anderson BDO, Kosut RL, Mareels IMY (1993) A new approach to adaptive robust control. International Journal of Adaptive Control and Signal Processing 7,3:183–211
Lewis FL (1992) Applied optimal control and estimation. Prentice Hall
Lipták BG (1995) Instrument engineers' handbook: Process control, 3rd ed. CRC Press
Ljung L (1999) System identification: Theory for the user. Prentice Hall, Upper Saddle River, NJ
Lotka AJ (1998) Analytical theory of biological populations. Plenum Press, New York
Lyapunov A (1992) The general problem of the stability of motion. Taylor and Francis, London

Maciejowski JM (2002) Predictive control: With constraints. Prentice Hall, Harlow
Mallat S (2001) A wavelet tour of signal processing, 2nd ed. Academic Press, San Diego
Mareels IMY, Polderman JW (1994) Adaptive systems: An introduction. Birkhäuser, Boston Basel Berlin
Mareels IMY, Weyer E, Ooi SK, Cantoni M, Li Y, Nair G (2005) Systems engineering for irrigation systems: Successes and challenges. Annual Reviews in Control 29(2):191–204
Matic N (2003) Introduction to PLC controllers. mikroElektronika, *http://www.mikroe.com/en/books/plcbook/*
Mayr O (1970) The origins of feedback control. MIT Press
McClellan JH, Schafer RW, Yoder MA (2003) Signal processing first. Pearson, Prentice Hall, Upper Saddle River
McMillan, GK, Cameron RA (2005) Advanced pH measurement and control, 3rd ed. International Society of Automation
Mees A (1991) Dynamics of feedback systems. John Wiley & Sons, Chichester, England
Mindell DA (2004) Between human and machine: Feedback, control, and computing before cybernetics. JHU Press
Mitra S (2005) Digital signal processing, 3rd ed. Mcgraw-Hill Science, New York
Moor J (2003) The Turing test: The elusive standard of artificial intelligence. Springer-Verlag, Berlin Heidelberg
Nijmeijer H, van der Schaft A (1990) Nonlinear dynamical control systems. Springer-Verlag, New York
Nyquist H (1932) Regeneration theory. Bell System Technical Journal 11:126–147
Oppenheim AV, Willsky AS, Hamid Nawab S (1982) Signals and systems. Prentice-Hall, New York
Ortega R, Loría A, Nicklasson PR, Sira-Ramirez HJ (1998) Passivity-based control of Euler-Lagrange systems: Mechanical, electrical, and electromechanical applications. Springer-Verlag, Berlin Heidelberg
Polderman JW, Willems JC (1998) Introduction to mathematical systems theory: A behavioral approach. Springer-Verlag, London
Poor VH (1994) An introduction to signal detection and estimation, 2nd ed. Springer-Verlag, New York Berlin
Pratt RW (ed) (2000) Flight control systems: Practial issues in design and implementation. IEE, Control Engineering Series 57
Ramirez M, Albertos P (2008) PID control with fuzzy adaptation of a metallurgical furnace. In: Granular computing: At the junction of rough sets and fuzzy sets, vol. 224. Springer-Verlag, Berlin Heidelberg
Remagnino P, Foresti GL, Ellis T (2005) Ambient intelligence: A novel paradigm. Springer-Verlag, Berlin Heidelberg
Roberts GW (2009) Chemical reactions and chemical reactors. John Wiley & Sons, Hoboken, NJ
Rohner P (1996) PLC: Automation with programmable logic controllers: A textbook for engineers and technicians. UNSW Press, Sydney
Roth RM (2006) Introduction to coding theory. Cambridge University Press, Cambidge
Santoro A, Ferramosca E, Mancini E (2008) Biofeedback-driven dialysis: Where are we? Contributions to Nephrology 161:199–209
Santoso L (2003) Managing type 1 diabetes mellitus: Modelling and control. The University of Melbourne, Melbourne, PhD Thesis
Santoso L, Mareels IMY (2002) A direct adaptive control strategy for managing diabetes mellitus. Proceedings of the IEEE Conference on Decision and Control, vol 3. Las Vegas, Nevada, pp 2530–2535
Schoukens J, Pintelon R (1991) Identification of linear systems: A practical guideline to accurate modeling. Pergamon Press, Oxford
Shannon CE (1949) The mathematical theory of communication. University of Illinois Press, Urbana
Shaw WT (2006) Cybersecurity for Scada systems. PennWell Books
Shinskey FG (1996) Process control systems: Application, design, and tuning, 4th ed. McGraw-Hill, New York
Shoham Y, Leyton-Brown K (2009) Multiagent systems algorithmic, game-theoretic, and logical foundations. Cambridge University Press
Siljak DD (1991) Decentralized control of complex systems. Academic Press, Boston
Skogestad S, Postlethwaite I (1996) Multivariable feedback control. Analysis and Design. John Wiley & Sons

Snyder SD (2000) Active noise control primer. Springer-Verlag, Berlin Heidelberg
Sontag E (1998) Mathematical control theory, deterministic finite dimensional systems, 2nd ed. Springer-Verlag, Texts in Applied Mathematics, Berlin
Soper ME (2005) Absolute beginner's guide to home automation. Macmillan Computer Pub
Stark H, Woods JW (2002) Probability and random processes with applications to signal processing, 3rd ed. Prentice Hall, Upper Saddle River
Sternby J (1996) Adaptive control in ultrafiltration. IEEE Trans. on Control System Technology 4: 1267–1281
Sydenham PH (1984) Transducers in measurement and control, 3rd ed. CRC press
The Australian Academy of Technology Science and Engineering (ed) (2002) Report of a Western Australia division study. The Australian Academy of Technology Science and Engineering, http://www.atse.org.au/publications/reports/water2.htm
Turing AM (1992) Mechanical intelligence. Edited by Ince DC, North Holland, Amsterdam New York
UNESCO (2006) The 2nd United Nations world water development report: Water; a shared responsibility. UNESCO and Berghahn Books, Barcelona, http://www.unesco.org/water/wwap/wwdr/
UNESCO (2009) The 3rd United Nations world water development report: Water in a changing world. UNESCO publishing and Beghahn Books, Barcelona, http://www.unesco.org/water/wwap/wwdr/
Vahid F, Givargis T (2001) Embedded system design: A unified hardware/software introduction. John Wiley & Sons
van der Schaft A (2000) L2-Gain and passivity techniques in nonlinear control, 2nd ed. Springer-Verlag, Berlin Heidelberg
Vidyasagar M (1996) A theory of learning and generalization with applications to neural networks and control systems. Springer-Verlag, Berlin Heidelberg
von Bartalanffy L (1980) General system theory: Foundations, development, applications. George Braziller
Von Neumann J (1958) The computer and the brain. Yale University Press, New Haven
Waldhauer FD (1982) Feedback. John Wiley & Sons
Weber W, Rabaey JM, Aarts EHL (2005) Ambient intelligence. Springer-Verlag, Berlin Heidelberg
Wells C (1996) The Kalman filter in finance. Kluwer Academic Publishers, Dordrecht Boston
Weyer E (2001) System identification of an open water channel. Control Engineering Practice 9(12): 1289–1299, www.scopus.com
Whitaker HP (1959) An adaptive system for control of the dynamics performance of aircraft and spacecraft. Inst. Aeronautical Sciences, Paper 59-100
Wiener N (1948) Cybernetics: On communication and control in animals and machines. MIT Press, Cambridge, MA
Wiener N (1954) The human use of human beings: Cybernetics and society, 2nd ed. Doubleday, New York
Wiener N (1961) Cybernetics: Or control and communication in the animal and the machine. MIT Press, Cambridge, MA
Wiggins S (2003) Introduction to applied nonlinear dynamical systems and chaos. Springer-Verlag, New York
Willems JC (1997) On interconnections, control and feedback. IEEE Transactions on Automatic Control 42:326–339
Willems JL (1970) Stability theory of dynamical systems. Nelson, London
Wilson DR (1969) Modern practice in servo design. Pergamon Press, New York
Wonham WM (1979) Linear multivariable control: A geometric approach. Springer-Verlag, New York
Zajonc RB (1984) On the primacy of affect. American Psychologist 35:151–175
Zeh HD (1992) The physical basis of the direction of time, 2nd ed. Springer-Verlag, Berlin
Zhou M, DiCesare F (1993) Petri net synthesis for discrete event control of manufacturing systems. Springer-Verlag, Berlin Heidelberg

Index

Symbols

2DoF control 216

A

active noise control 279, 287
actuator 23, 241
 –, dual 243
 –, nonlinearities 242
 –, smart 243
adaptation 208, 281
adaptive control 208, 263
 –, gain scheduling 266
 –, MIT rule 84, 264
 –, model reference adaptive system (MRAS) 264
 –, self tuning 265
ADC (see *Analog to Digital Converter*)
agent 301
aliasing 104
ambient intelligence 300
amplifier, operational 184
Analog to Digital Converter (ADC) 228, 233
analogy 36, 44
antenna control 65, 216
aperiodic 98
Aristotle 227
artificial intelligence 298
attractivity 155
automata 71, 211, 236

B

bandwidth 100, 138
behavior 17
 –, analogy 43
bifurcation 159, 171
 –, diagram 162
 –, theory 159
bio-component system 304
biomimetics 303
bionics 279, 303
block diagram 5
 –, calculus 133
Bode, Hendrik 136
Bode plot 137
boiler 212
Boole, George 120
Boolean algebra 256

C

calibration 98
cascade 22
 –, stability of 169
causality 4, 13, 18, 123
CCC (see *communication, computation and control*)
chaos 171, 172
characteristic equation 157
Clark, Graeme 287
coding 106
communication 220, 235
 –, network 228
communication, computation and control (CCC) 220, 245, 296
complex number 92
complexity 280
compression 107
conditionally stable 189
continuity 151
control
 –, cascade 69, 81, 216
 –, closed-loop 80, 214
 –, codesign 241
 –, computer based 239
 –, coordination 209
 –, decoupling 262
 –, design 142, 253
  –, process 253
 –, distributed 59, 219

control (*continued*)
 –, feed-forward 195, 212
 –, fundamental limits 251
 –, goals 207
 –, hierarchical 57, 82, 219, 220
 –, hierarchy 59
 –, hybrid 273
 –, interface 207
 –, learning 209, 267
 –, levels 252
 –, logic 255, 256
 –, Model Predictive (MPC) 272
 –, multivariable 212
 –, network 220
 –, on-off 237
 –, open-loop 79, 205, 210, 213
 –, optimal 270
 –, P- 238
 –, PD- 238
 –, performance 195, 249, 254
 –, PI - 238
 –, PID- 186, 238, 239
 –, process redesign 221
 –, quality 213
 –, robust 271
 –, selective 82, 217
 –, servo 65
  –, -system 69
 –, supervisory 70, 208, 269
 –, tracking 68, 70, 186, 257
 –, two-degrees-of-freedom 196, 216
controller 23
cyber-physical systems 295

D

DAC (see *Digital to Analog Converter*)
DAS (see *Data Acquisition System*)
Data Acquisition System (DAS) 23, 206, 227, 228, 230, 231
Data Signal Processor (DSP) 23, 228
data to model transform 25, 117
dead beat control 161
decentralized 220
decoupling 262
delay 13
design, integrated 221
Digital to Analog Converter (DAC) 228, 233
disturbance 175
 –, rejection 193, 207
domotics 279
DSP (see *Data Signal Processor*)
dual actuator 243

E

efficiency 280, 282
eigenvalue 156
Einstein, Albert 115
embedded
 –, control system 292
 –, system 232
engineered system 249
equilibrium 15, 41, 154
 –, stable 154
 –, unstable 154
Euler, Leonhard 91
expert system 298
exponential signal 91

F

fault detection 208
federated control 294
feed forward 80
feedback 3–5, 13, 133, 214
 –, acoustic 14
 –, control 205
 –, internal 182
 –, loop 4, 122
 –, negative 15, 45, 184
 –, positive 15
 –, stabilization 186
 –, state 215
 –, tracking 186
feedforward 212
filter 194
 –, Kalman 234
Fourier, Jean Baptiste 88
Fourier
 –, series 97
 –, transform 100
frequency
 –, domain 136
 –, response 136
Friedman, Milton 53
function 32
fuzzy logic 299

G

gain 125
 –, margin 189
 –, scheduling 266
Gauss, Johann Carl Friedrich 118
Gibbs, Josiah Willard 227
governor 207, 291

H

harmonic analysis 138
hemodialysis 287
Hertz, Heinrich 95
heuristics 258
hierarchical control 220
home automation 279
homeostasis 72
Hunter, Peter 302
Hurwitz, Adolf 204

I

information
 -, flow 206
 -, loop 4
input 5
input-to-state stability (ISS) 167–169, 188
instability 155
integral action 186
integrated design 221
integrator 128
 -, examples 128
 -, in feedback 130
 -, notation 128
 -, properties 129
interpolation 43
invariance 124
inverse problem 142
irrigation system 59
ISS (see *input-to-state stability*)

J

Joint Photographic Experts Group (JPEG) 88
junction, neuromuscular 207

K

Kalman, Rudolph 234
Kalman filter 234
Kepler, Johannes 20
kiln 57

L

Landau, Ioan Dore 265
Laplace, Pierre-Simon 100
Laplace transform 100, 101, 131, 158
learning 209, 267, 281
least squares 118
Lebesgue, Henri 103

linear 18
Linear Time-Invariant (LTI) system 123
linearity 124
linearization 120
logic control 255, 256
loop-gain 188
LTI system (see *Linear Time-Invariant system*)
Lyapunov, Aleksandr 163
Lyapunov
 -, function 165
 -, method
   -, first 163
   -, second 165

M

machine learning 298
manufacturing 235
 -, system 54, 213, 219
master-slave 211, 298
Maxwell, James 15, 204
measurement, noise 194
mechanics 154
MIMO system (see *Multi-Input-Multi-Output system*)
model 9, 19, 116, 121
 -, discrete time 144
 -, input-output 119
 -, limits 152
 -, Lnon linear 145
 -, logic 119
 -, state space 119, 143
 -, uncertainty 174
Model Predictive Control (MPC) 272
Model Reference Adaptive System (MRAS) 264
modeling 21
motor control 207
Moving Picture Experts Group (MPEG) 88
MP3 88, 111, 303
MPC (see *Model Predictive Control*)
MPEG (see *Moving Picture Experts Group*)
MRAS (see *Model Reference Adaptive System*)
multi-agent system 301
Multi-Input-Multi-Output (MIMO) system 261, 262
multivariable control 212

N

networked control 220
 -, system 294
Neumann, von, John 71

neural network 299
Newton, Isaac 20
Nichols, Nathaniel 258
noise measurement 194
Nyquist, Harry 104, 191
Nyquist plot 137

O

observability 234
open loop 24, 211
–, control 79, 205, 210, 213
operational amplifier 184
operator 8
optimal control 270
optimization 208
output 5

P

P-control 238
parallel connection 133
Pavlov, Ivan 79
PD-control 238
period 87, 95
pH regulation 221
PI-control 238
PID (Proportional, Integral and Derivative action) 186, 238
PID-control 238, 239
Plancherel, Michel 100
PLC (see *Programmable Logic Controller*)
pole 156
preventive maintenance 281
process coupling 243, 261
Programmable Logic Controller (PLC) 236, 256

Q

quality control 213
quasi periodic 98

R

random 98
–, signal 98
–, variable 98
real time 240
receiver 115
regulation 186, 207, 257
relay 237
reliability 281

repetitive learning control 268
resonance 138, 139
robust
–, performance 175
–, stability 175
robustness 151, 174
Routh, Edward. J. 204

S

safety 281
safety, reliability and longevity 280
sampling 43
–, criterion 105
SCADA (see *Supervisory Control And Data Acquisition*)
scheduling 70
sensitivity 151, 172, 176
–, measurement 177
sensor 230
–, –/actuator network 220
–, network 235
–, smart 232
–, soft 231, 233
–, system 232
–, virtual 233
senstivity
–, complementary 176
sequencing 208, 210
series connection 133
servo control 65
Shannon, Claude 106
signal 8, 32, 34, 35, 88
–, aliasing 104
–, analogy 38
–, aperiodic 98
–, chaos 99
–, coding 106
–, compression 107
–, continuous time 35
–, deterministic 96
–, discrete 35
–, energy 100
–, exponential 91
–, external 12
–, Fourier 97
–, generator 115
–, harmonic 97
–, hybrid 229
–, index 34
–, input 12
–, internal 12
–, Laplace transform 100

-, output 12
-, periodic 95
-, processing 35
-, quasi periodic 98
-, random 98
-, sampled data 35
-, sampling 104
-, sine 94
-, size 102
-, stochastic 98
-, transform 100
-, value 34
-, Z-transform 100, 145
simulation 147
sinusoid 93
small gain theorem 188
smart
   -, actuator 243
   -, sensor 232
soft sensor 231, 233
spectrum 100
stability 151, 154, 155
   -, asymptotic 155
   -, BIBO 169
   -, gain margin 189
   -, input-to-state 167, 188
   -, limit cycle 171
   -, linear systems 156
   -, Lyapunov 163
   -, non linear systems 163, 167
   -, Nyquist criterion 191
   -, robust 175
state 19, 144, 167, 215
   -, feedback 215
stationarity 18
stationary 123
stochastic 98
subsystem 5
supervision 208, 269
supervisory control 209
Supervisory Control And Data Acquisition (SCADA) 60, 103, 235, 286, 287
system 5
   -, analogy 44
   -, analysis 135
   -, autonomous 154
   -, behavior 116
   -, cascade 91, 139
   -, causal 123
   -, complex 54
   -, delay 127
   -, eigenvalue 156
   -, embedded 232, 292

-, expert s. 298
-, frequency response 132
-, hybrid 229
-, identification 117
-, integrator 128
-, interconnection 22, 121, 169, 205
-, inverse response 218
-, linear 124, 130
-, Linear Time-Invariant (LTI) 123
-, manufacturing 54, 213, 219
-, model 116, 121
-, Model Reference Adaptive (MRA) 264
-, multi-agent 301
-, Multi-Input-Multi-Output (MIMO) 261, 262
-, non-minimum phase 218
-, oscillatory 47
-, pole 156
-, reactive 182
-, realization 142
-, response 135
-, social 76
-, stable 152, 169
-, state 19, 143
-, synthesis 140
-, time invariant 123
-, to model 118
-, transfer function 131
-, unstable 153, 169

T

tele-operation 211, 298
thermodynamics 39
   -, fist law 39
time
   -, invariance 123
   -, response 135
   -, series 32
tracking 208
transducer 207, 230, 232, 241
transfer function 131, 132, 145
transform
   -, data to model 25, 117
   -, Fourier 100
   -, Laplace 100, 101, 131, 158
   -, wavelet 112
   -, Z- 100, 145, 157
transmitter 115
transport delay 127
Turing, Alan 71, 297, 304
Turing test 297

U

uncertainty
 –, parametric 175
 –, structural 175
 –, unstructured 175

V

variable 6, 9, 21, 32
 –, free 9, 121
 –, random 98

W

Watt, James 203
wavelet transform 112
Weierstrass, Karl 90
Wiener, Norbert 23, 296

Z

Z-transform 100, 145, 157
Ziegler, John 258
Ziegler-Nichols 258